Wind Tunnel Testing of High-Rise Buildings

An output of the CTBUH Wind Engineering Working Group

Peter Irwin, Roy Denoon & David Scott

NEW YORK AND LONDON

Bibliographic Reference:
Irwin, P., Denoon, R. & Scott, D. (2013) *Wind Tunnel Testing of High-Rise Buildings: An output of the CTBUH Wind Engineering Working Group.* Council on Tall Buildings and Urban Habitat: Chicago.

Principal Authors: Peter Irwin, Roy Denoon & David Scott
Layout & Design: Steven Henry

First published 2013 by Routledge
2 Park Square, Milton Park, Abingdon, Oxon, OX14 4RN

Simultaneously published in the USA and Canada by Routledge
711 Third Avenue, New York, NY 10017

Routledge is an imprint of the Taylor & Francis Group, an informa business

Published in conjunction with the Council on Tall Buildings and Urban Habitat (CTBUH) and the Illinois Institute of Technology

British Library Cataloguing in Publication Data
A catalogue record for this book is available from the British Library

Library of Congress Cataloging-in-Publication Data
A catalog record has been requested for this book

ISBN13 978-0-415714-59-4

Council on Tall Buildings and Urban Habitat
S.R. Crown Hall
Illinois Institute of Technology
3360 South State Street
Chicago, IL 60616
Phone: +1 (312) 567-3487
Fax: +1 (312) 567-3820
Email: info@ctbuh.org
http://www.ctbuh.org

Front Cover: Greenland Center, Wuhan China. Under construction (2017 – expected completion).
© Adrian Smith + Gordon Gill Architecture

Principal Authors

Peter Irwin, RWDI, Canada
Roy Denoon, CPP, USA
David Scott, Laing O'Rourke, UK

Peer Review Panel

William Baker, SOM, USA
Melissa Burton, BMT, UK
Alberto Fainstein, AHFsa Consulting Engineers, Argentina
Jon Galsworthy, RWDI, Canada
Dr. Yaojun Ge, Tongjii University, China
T.C. Eric Ho, Boundary Layer Wind Tunnel Laboratory, Canada
Alan Jalil, Jacobs, France
Prof. Sang Dae Kim, Korea University, Korea
Prof. Acir Loredo-Souza, Universidade Federal do Rio Grande do Sul, Brazil
Richard Marshall, Buro Happold, UAE
Bill Melbourne, MEL Consultants, Australia
Tony Rofail, Windtech Consultants Pty., Ltd., Australia
Bob Sinn, Thornton Tomasetti, USA
Wolfgang Sundermann, Werner Sobek Group, Germany
Prof. Yukio Tamura, Tokyo Polytechnic University, Japan

Contents

About the CTBUH

The Council on Tall Buildings and Urban Habitat is the world's leading resource for professionals focused on the design, construction, and operation of tall buildings and future cities. A not-for-profit organization based at the Illinois Institute of Technology, Chicago, the group facilitates the exchange of the latest knowledge available on tall buildings around the world through events, publications, research, working groups, web resources, and its extensive network of international representatives. Its free database on tall buildings, The Skyscraper Center, is updated daily with detailed information, images, data, and news. The CTBUH also developed the international standards for measuring tall building height and is recognized as the arbiter for bestowing such designations as "The World's Tallest Building."

About the Authors

Peter Irwin
RWDI, Canada

A Principal of Rowan, Williams, Davies and Irwin, Inc., Peter joined the company in 1980 and served as president from 1999–2008. His experience in wind engineering dates back to 1974 and includes extensive research and consulting in wind loading, aeroelastic response, wind tunnel methods and instrumentation, as well as supervising many hundreds of wind engineering studies of major structures. He has worked on the Petronas Towers, Kuala Lumpur; Taipei 101, Taiwan; Two International Finance Centre, Hong Kong; and the Burj Khalifa, Dubai.

Roy Denoon
CPP, USA

Roy joined CPP in 2004 following several years working in Hong Kong, where he was responsible for the wind engineering of such iconic tall buildings as Union Square in Hong Kong and the CCTV Headquarters in Beijing. Since joining CPP, Roy has been responsible for numerous tall building projects throughout the world and now leads CPP's structural wind engineering group. He recently supervised the instrumentation of Burj Khalifa in Dubai to monitor its long-term dynamic response characteristics and is enjoying analyzing the results.

David Scott
Laing O'Rourke, UK

Currently the structural director of the Engineering Excellence Team at Laing O'Rourke, David is also a past Chairman of the CTBUH (2006–2009), and led Arup's business practice in the Americas for many years. He was at the forefront of many of Arup's biggest projects, including work on the Hong Kong and Shanghai Bank and Cheung Kong Center. He also spent 15 years based in southeast Asia, where he was involved with Arup's first major projects in China, Korea, Indonesia, and the Philippines.

Preface

Since the 1960s, wind tunnel testing has become a commonly used tool in the design of tall buildings. It was pioneered, in large part, during the design of the World Trade Center Towers in New York. Since those early days of wind engineering, wind tunnel testing techniques have developed in sophistication, but these techniques are not widely understood by the designers using the results. The CTBUH recognized the need to improve understanding of wind tunnel testing in the design community. The CTBUH Wind Engineering Working Group was formed to develop a concise guide for the non-specialist.

Objectives of this Guide

The primary goal of this guide is to provide an overview of the wind tunnel testing process for design professionals. This knowledge should allow readers to ask the correct questions of their wind engineering consultants throughout the design process. The guide is not intended to be an in-depth guide to the technical intricacies of wind tunnel testing, as these are covered in several other publications. The guide does, however, introduce one topic that has not been addressed previously, but which the design community needs: a methodology for the presentation of wind tunnel results to allow straightforward comparison of results from different wind tunnel laboratories. The wind loads provided by wind engineering specialists have a major effect on the construction costs of many tall buildings. Parallel wind tunnel tests by different laboratories are becoming more frequent, either as part of a peer review process, or as a more direct attempt to reduce design loads. The loads provided by different wind engineering consultants are never identical, and can sometimes be markedly different. The framework presented here is specifically designed to facilitate comparisons of results and to allow the identification of the sources of any differences.

Content Overview

Wind tunnel testing is used in the design of most major tall buildings to identify the wind-induced structural loads and responses for which the superstructure must be designed. The processes by which wind engineers predict these loads and responses can appear arcane to many of the designers who have to use the results. This can, in some cases, lead to a reluctance to rely on wind tunnel predictions. This guide is intended to shed light on the science of wind engineering and the derivation of the conclusions provided in wind tunnel test reports.

The first wind engineering task for many designers is to determine whether to design using a local wind loading code or standard, or whether to employ wind tunnel testing. This guide begins with basic advice on when a tall building is likely to be sufficiently sensitive to wind effects to benefit from a wind tunnel test and provides background for assessing whether design codes and standards are applicable.

Once a decision to proceed with wind tunnel testing has been reached, it is important to ensure that the appropriate tests are being specified and conducted. In addition to providing details of the types of tests that are commonly conducted, descriptions of the fundamentals of wind climate and the interaction of wind and tall buildings is provided in order for the reader to be able to put the use of wind tunnel tests into context. While the majority of this guide concentrates on the testing that is conducted to determine the overall structural loads and responses, brief descriptions of other studies that may be conducted during design are also provided to alert the reader to aspects of wind engineering that may be relevant to particular design features.

Different laboratories use different techniques to combine the basic loads measured in the wind tunnel with the statistical descriptions of wind climate that are necessary to provide loads and responses with a known probability of exceedance, or which are consistent with a specified return period. This can be one of the largest causes of differences in results from different laboratories. In this guide, these different approaches are explained clearly, and the advantages and disadvantages of each summarized. Understanding that results from different laboratories may be different, it is important for a design team to identify the sources of such differences. This guide provides a standardized results presentation format to facilitate comparison. This provides a straightforward method for a design team to assess whether differences are due to factors such as fundamentally different aerodynamic characteristics being measured in the wind tunnel, or different conclusions having been reached in the wind climate analysis, or different methods having been used to combine the wind climate and aerodynamic coefficients. This knowledge then affords educated design decisions regarding wind loads.

The Wind Engineering Working Group hopes this guide is useful to the design professionals for whom it is intended and welcomes any feedback that can be used to improve future editions.

1.0 Introduction

The main structure of a tall building and its façade must be designed to safely withstand the extreme winds to which the building will be subjected during its expected life. Determining what the wind loads will be for specific mean recurrence intervals, and what the uncertainties are in these loads is critical. The wind loads, and appropriate load factors that allow for uncertainty in ordinary buildings, are often prescribed by the analytical methods given in building codes. But for tall buildings, in view of the importance of wind loads to their cost and safety, these analytical methods often lack the precision needed. Also, they do not account well for important wind phenomena, such as crosswind excitation, aerodynamic interactions between adjacent buildings, and aerodynamic instability, all of which affect not only loads but may also cause building motions that occupants find excessive. For these reasons, the wind loads and motions of tall buildings are typically determined by wind tunnel tests on scale models of the building and its surroundings, through which much more precise, project-specific information is obtained. Computational Fluid Dynamics (CFD) is increasingly used for qualitative evaluation of wind effects, particularly ground-level wind speeds, but is not yet capable of providing quantitative results of sufficient accuracy for the determination of design wind loads.

The objective of this document is to lay out general guidelines for wind tunnel tests, as applied to tall buildings, in a format that is useful to building professionals and regulatory authorities involved in tall buildings, as well as wind specialists. It is not intended to be a detailed manual of practice, such as is provided in: ASCE, 1999; AWES, 2001; BCJ, 1993 & 2008; KCTBUH, 2009; and ASCE 49–12, 2012. However, it is intended to describe best practice and make it easier to compare results from different wind consultants.

1.1 Basis of Design

Wind tunnel testing involves highly developed and specialized methodologies and terminology. Designers, developers, and building officials cannot be expected to have the in-depth knowledge of such a specialized field but it will help them to obtain most value from wind tunnel tests of their projects if they have a basic understanding of the principles involved. Also, as with any branch of knowledge, it is important to be aware of the sources of uncertainty in wind studies so that proper judgement can be exercised when applying the results or comparing results from different wind tunnel test laboratories.

The wind load formulae of building codes have been developed primarily for low-rise buildings and typically address, with a few exceptions, only wind loads in the along-wind direction. They are specified as the product of various factors, such as a reference pressure q, an exposure factor k, a drag coefficient

> **The wind load formulae of building codes have been developed primarily for low-rise buildings and typically address only wind loads in the along-wind direction.**

▲ Figure 1.1: Tall buildings designed for dense urban settings such as Chicago (above) benefit from the precise, project-specific information obtained from wind tunnel testing.

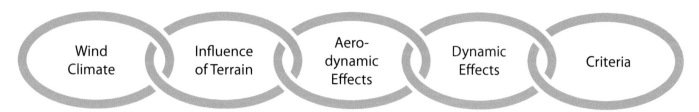

▲ Figure 1.2: The Alan G. Davenport Chain of Wind Loading.

Cd, and a gust factor Cg. This has sometimes led to the expectation that the purpose of the wind tunnel test is simply to determine the drag coefficient and possibly the gust factor, the values of which are then to be inserted into the formula. This expectation misses the important point that, for buildings dominated by crosswind loading, the format of the typical code formula does not capture the essential physics of the problem. The objective of the wind tunnel tests is to fully replicate the real physics of wind loading at model scale. This includes along-wind loading, crosswind loading, torsional loading, load combinations, building motions, local wind pressures for cladding design, and the influences of terrain roughness, topography, directionality, and other nearby structures (see Figure 1.1).

When does a building's height make it sufficiently sensitive to dynamic effects and crosswind loading to require a wind tunnel test? The answer depends on many factors, including its shape, exposure, slenderness, structural system, and the wind regime of the site location. A wind tunnel test may be advisable if any one of the following applies:

(i) The height of the building is over 120 meters.
(ii) The height of the building is greater than four times its average b_{av} (width normal to the wind direction over the top half of the building).
(iii) The lowest natural frequency of the building is less than 0.25 Hz.

(iv) The reduced velocity $U / (f_1 b_{av})$ at ultimate conditions is greater than five, where U is the mean hourly wind velocity evaluated at the top of the building, f_1 is the lowest natural frequency of the building and b_{av} is the average width defined in (ii).

It should be noted that these are approximate guidelines only, and can depend on other factors, such as exposure of the building being considered, local topography, and the presence of other major buildings in the proximity.

There are five key steps involved in determining wind loadings derived from wind tunnel tests. This has been described as a "chain" (Davenport 1982), which is appropriate, since the outcome is only as strong as the weakest link. The Alan G. Davenport Chain is illustrated in Figure 1.2.

The first link in the chain is the wind climate, i.e., the statistics of the wind speed and direction for the region where the building is located. The next link is the influence of the surrounding terrain, including the surface roughness and topography (see Figure 1.3). This is followed by a link representing the local aerodynamics of the building and interference effects from other nearby structures. The next link, dynamic effects, represents the building's wind-induced response, including any aeroelastic effects. The final link represents the criteria used to assess the building and its response to wind. The objective of

present-day wind tunnel studies is to evaluate each link in detail using rational methods and with maximum accuracy. Then, when all links are assembled, the final answer is the best available using rational scientific knowledge. If one or more of the links is not treated with due diligence, the value of the whole study can be seriously degraded. For this reason, it is usually not a good idea to break off one part of the chain of the wind tunnel study and substitute part of the code analytical procedure, which is inherently more approximate, in its place. The analytical procedure of the code and the wind tunnel procedure are best treated as two totally separate processes, each targeted at the same level of structural reliability, and only their end results compared.

1.2 Wind Climate

Wind climate involves the statistics of wind speed and direction. Building codes generally specify the recurrence interval and the corresponding wind speed at a selected reference height, typically at 10 meters in open terrain. Some codes go so far as to specify different wind speeds for different wind directions. Depending on the country involved and the location within that country, more or less effort will have gone into the determination of the appropriate design wind speed. Where the design wind speed is based on well documented research by established experts, it would be normal for the wind

▲ Figure 1.3: Topography model test of Hong Kong. © CPP

tunnel laboratory to make predictions based on a statistical wind model that matches that speed. However, tall buildings are sometimes built in areas where the research resources have not been available to determine an accurate design speed and there has been little prior experience with tall structures. In such cases the determination of design wind speed made by the wind tunnel laboratory's experts may well be more reliable and rational than the code speed. Indeed, there are some jurisdictions where the code wind speed has been updated based on the results of studies done for a specific tall building. The fact is that tall buildings are extremely sensitive to wind speed and direction. Therefore, every effort should be made to determine the statistics of these parameters in the most rational manner possible.

The first stage of any wind engineering study is to conduct a wind climate analysis for the development site. The most common sources of wind data are anemometer records from local

meteorological stations, usually located at airports. These data are often available for each hour of the record period, but at some locations may not be available at such frequent intervals, or there may be gaps in the record. While the ideal anemometer location would be surrounded by flat, open terrain in all directions, this ideal situation is typically not realized. Therefore the records need to be corrected back to the standard "open-country" condition to make them comparable to other locations and to code-defined wind speeds. This is typically done using a methodology such as that published by ESDU (1993), which takes account of the effects of terrain changes on wind characteristics. This same methodology can then be used to adjust the open-country data to site wind conditions, taking account of the roughness upwind from the site for each approach direction.

The number of years for which meteorological records are available is important in obtaining statistical reliability. A 15-year or longer period is desirable.

The existing practice is to assume that the statistics of the past wind climate give a good indication of the future. This practice has been questioned in recent years on account of climate change predictions, but currently the uncertainties in predicting the probable effects on extreme winds have resulted in no clearly identifiable trend. If and when clear trends are predicted, existing practice may need to be modified. Apart from their statistical reliability, interpretation of wind records is subject to other complicating factors.

Anemometers are not infrequently moved to a different location and height, and the roughness of the terrain surrounding them may have changed during the period of record. Also, recording errors have been known to degrade confidence in the data. These effects need to be taken into account as much as possible, but clearly they introduce added uncertainty beyond that coming from pure length of record. The uncertainties connected with individual meteorological stations can be reduced by examining more than one station in a given region. This is sometimes done using the "superstation" concept (Peterka & Shahid 1998), in which records from a number of nearby meteorological stations are combined into a single record, effectively creating a longer record length and an increased statistical reliability.

An important parameter for meteorological studies is the averaging time used for recording the anemometer data. Various averaging times are used, ranging from one hour to one minute. This needs to be taken into account in the analysis of data. Also, at some meteorological stations, peak daily gusts are recorded with durations on the order of one to three seconds. These can be useful in cross-checking the hourly data, and may also provide additional information on the terrain roughness and short-duration wind events, such as thunderstorms.

Since meteorological station anemometers are typically located at a height of 10 meters, it is clearly a considerable extrapolation to infer from them the wind statistics at the tops of tall buildings, some of which are now encroaching on the one-kilometer mark. Experience indicates that current extrapolation methods, using standard models of the planetary boundary layer, work reasonably well in practice for buildings up to 300 or 400 meters. However, it is desirable to obtain more direct information on winds at altitude, i.e., above 400 meters. At some locations, many fewer than those with ground-based stations, this is available in the form of balloon records. Unfortunately they are often only available at three-hour intervals at best, and more often than not, only at six-hour or 12-hour intervals, and frequently have missing records due to balloon malfunction or other reasons. However, provided one recognizes the uncertainty caused by the sparseness of the data set, baloon stations can shed some additional light on upper-level winds.

Another source is the United States' National Center for Atmospheric Research / National Center for Environmental Prediction (NCAR/NCEP) global reanalysis data set. These data are based on worldwide meteorological observations interpolated to a three-dimensional grid by means of meteorological modeling. The NCAR/NCEP global reanalysis dataset has been generated by the weather forecasting community. These data are now available on a worldwide grid at multiple levels, with spatial resolutions varying from 12 kilometers in some parts of the world to 200 kilometers in others, and temporal resolutions in the 3–6 hour range. The spatial and temporal resolutions can be enhanced locally, using specialized computational tools such as the Weather Research and Forecasting (WRF) tools developed by NCAR/NCEP and other agencies. For large supertall projects the use of WRF software, and its predecessor software MM5, has been used to improve understanding of upper-level winds (Qiu et al. 2005). Also, WRF studies are useful as an interpolation tool when the project site is located far from any meteorological station. It should be noted that these tools are in their early stages of use in tall building design.

Since the wind loading of tall buildings is often dominated by their dynamic response, and since they tend to have long natural periods of vibration, the duration of wind events important to these structures ranges from a few minutes to about an hour. Shorter events do not last long enough to generate a large dynamic response. Longer events can be treated as a series of periods, within each of which the statistics of the building's response do not change, i.e., they satisfy the stationarity criterion as defined in random vibration theory. Wind tunnel studies therefore typically assume statistically stationary conditions for periods ranging from 10 minutes to one hour. Large-scale storms, such as extra-tropical cyclones, hurricanes (see Figure 1.4), and typhoons certainly satisfy this type of stationarity assumption. However, because of their relatively short duration, thunderstorms and other local convective phenomena do not. A frequent cause of strong winds from thunderstorms is the downburst that causes a jet of air close to the ground. Because the maximum velocity occurs at a height well below the height of typical tall building, downbursts are unlikely to govern the overall structural loading, although cladding loads in the lower parts could be affected. Tornadoes are also spawned by thunderstorms, but their spatial extent is so small that the probability of a direct hit is generally very small and can be discounted when considering overall reliability. The wind

▲ Figure 1.4: Wind tunnel studies typically assume statistically stationary conditions for periods ranging from 10 minutes to one hour. Large-scale storms satisfy this type of stationarity assumption.

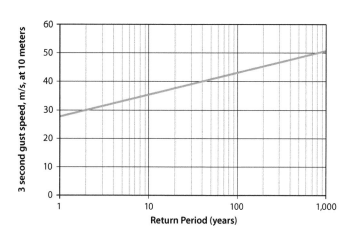

▲ Figure 1.5: Example of gust wind speed versus return period. Note: This plot is location-dependent.

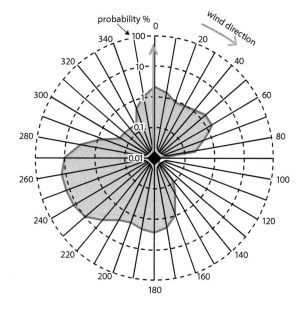

▲ Figure 1.6: Example of probability (in percent) of winds that are above the 50-year value being exceeded from each 10-degree sector.

Since the wind loading of tall buildings is often dominated by their dynamic response, and since they tend to have long natural periods of vibration, the duration of wind events important to these structures ranges from a few minutes to about an hour.

phenomena in thunderstorms are still the subject of ongoing research. Current practice in wind tunnel testing is to apply standard boundary layer profiles seen in large scale storms in the wind tunnel.

In parts of the world where extreme winds are dominated by hurricanes or typhoons, the sample of these storms recorded at any one meteorological station is usually too small for statistically reliable predictions to be made of extreme wind speeds and directions. This shortcoming has been to a large extent overcome through use of Monte Carlo simulation techniques, based on records of hurricanes and typhoons in the entire ocean region of concern rather than those from a single station. More details of these techniques are described by Vickery et al. 2010. It should be noted that in some countries such as the USA and Australia, the building code wind speeds along the coastlines affected by hurricanes and typhoons have been determined through Monte Carlo techniques. However, at the time

of writing, there are other areas where this is not the case, and the local code wind speeds may be affected by the uncertainties associated with the small number of these storms actually present in the data.

The methods of using meteorological data together with wind tunnel data are described in Section 4.0. Whichever method is used, a basic plot of wind speed versus return period at a chosen reference location and height should be generated as illustrated in Figure 1.5. Likewise an analysis illustrating the directional probability of wind should be generated as illustrated in Figure 1.6. In this example diagram, north is at zero degrees and the lobe in the plotted curve to the west-southwest illustrates that this direction has the highest probability of exceeding the 50-year wind speed. The radial scale in Figure 1.6 is logarithmic, because probabilities vary over a wide range of values.

2.0 Nature of Wind Effects

2.1 Planetary Boundary Layer and Wind Turbulence

The wind pressures acting on a tall building are very complex. While they clearly depend on the shape of the building itself, they also depend on the wind shear and turbulence caused by the roughness of the upwind terrain and the building's aerodynamic interaction with other nearby buildings. The earth's terrain roughness slows the wind near the ground, thus giving rise to the planetary boundary layer, within which the wind velocity increases with height (i.e., there is wind shear). Within this boundary layer the wind is also highly turbulent. The thickness of the planetary boundary layer can vary considerably, but at high wind speeds in synoptic storms it is typically two to three kilometers. In the eye-wall of hurricanes it tends to be less, in the 500-meter to 600-meter range, but further out from the eye-wall it ranges up to values similar to other large-scale storms. In localized thunderstorm gust fronts, the planetary boundary layer may only be 100 meters or so, above which the wind velocity drops off, thus making these gust fronts less critical to the overall loading of tall buildings of several hundred meters in height.

Because most of the changes in speed and turbulence in large-scale storms occur within the lowest few hundred meters, many building codes have, in the past, assumed for simplicity that above about 250 meters to 450 meters, depending on terrain, the wind speed remains constant. The impact of this approximation was minimal for buildings less than about 300 meters high, which until a few years ago was a height that was rarely exceeded. However, for establishing the wind loading on tall buildings in the 400-meter to 1,000-meter height range, which is becoming increasingly common, the full height of the planetary boundary layer in large-scale storms needs to be accounted for, and, as indicated above, it can range up to several kilometers at high wind speeds.

Wind turbulence in the planetary boundary layer is typically expressed by the longitudinal turbulence intensity, which is the ratio of the standard deviation of velocity fluctuations in the mean wind direction to the mean velocity. The turbulence intensity typically ranges from 10–30% near the earth's surface to 5–10% in the 500-meter to 1,000-meter height range, although much higher intensities are possible among groups of

tall buildings or in complex topography. Figure 2.1(a) illustrates the mean and instantaneous wind velocity profile approaching a tall building. The mean is averaged over about one hour and the instantaneous profile varies from instant to instant due to turbulence.

2.2 Mean and Fluctuating Loads

The wind pressures on a tall building fluctuate, not only because the oncoming wind is turbulent, but also because the building creates its own signature turbulence, Figure 2.1(b). The fluctuations due to both sources of turbulence occur much more rapidly than the changes in wind velocity, due to the passage of meteorological systems over the site. Changes in velocity due to the passage of large-scale meteorological systems occur over hours, whereas turbulence fluctuations occur over seconds. The latter are sensed by observers as gusts, whereas the former are sensed as gradual changes in the general magnitude of wind speed. The dividing line between the durations of turbulence events and large-scale meteorological events is usually set at about 10 minutes to an hour. Thus, it is conventional to describe as mean

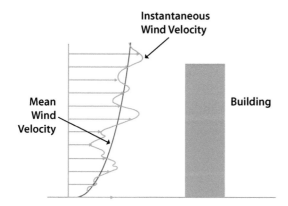

▲ Figure 2.1: Wind profile and wind turbulence: (a) Elevation of building in turbulent and sheared wind, (b) Plan view showing signature turbulence.

The wind pressures on a tall building fluctuate not only because the oncoming wind is turbulent, but also because the building creates its own signature turbulence. The fluctuations due to both sources of turbulence occur much more rapidly than the changes in wind velocity due to the passage of meteorological systems over the site.

loads those obtained by averaging over one hour. These are associated with the mean wind velocity and direction averaged over the hour. To characterize the fluctuations in load about the mean, researchers use statistical descriptions such as standard deviation of load and expected peak load within a one-hour period.

2.3 Along-wind, Crosswind, and Torsional Loading

It is important to note that buildings experience loads, not only in the direction of the wind but also at right angles to it, i.e., in the crosswind direction. Both mean and fluctuating crosswind loads will occur for buildings that lack symmetry, or where the surroundings cause asymmetrical flows. For a perfectly symmetrical building in surroundings that do not disturb the symmetry, the mean crosswind loads are indeed zero, but there are still substantial crosswind load fluctuations. These are due to fluctuations in both the lateral component of turbulence velocity in the approaching

wind and the building's own signature turbulence. It is usual for the crosswind loading to be of similar magnitude to, or higher than, the along-wind loading on tall, slender buildings, even though the along-wind loads include a large mean component. As with mean crosswind loading, mean torsional loading can occur when there is asymmetry in the building (architectural form or structural system) and/or surroundings. But even in a perfectly symmetrical case, torsion loading fluctuations will occur due to turbulence effects. Typically, peak along-wind, crosswind, and torsional loadings do not all occur at the same instant. It is important for wind tunnel studies to determine appropriate combinations of these loads.

2.4 Background and Resonant Loads

The fluctuations in wind load due to the direct action of the instantaneous wind pressures applied over all exterior surfaces of the building are called the background loading. In many cases, the background loading is not evenly distributed over the height of the building because of the chaotic nature of the oncoming turbulence in the wind. In this situation, the shape of the load distribution with height varies considerably from moment to moment, and there is little correlation between what is happening at the top and bottom of the building. The situation is altered when a building's shape promotes strong vortex excitation. In that case, the signature turbulence of the building itself dominates over the oncoming turbulence, leading to greatly enhanced correlation of the background loading up the height of the building, which can in turn lead to significantly higher crosswind loads.

The sustained effect of the background load fluctuations on the building is to

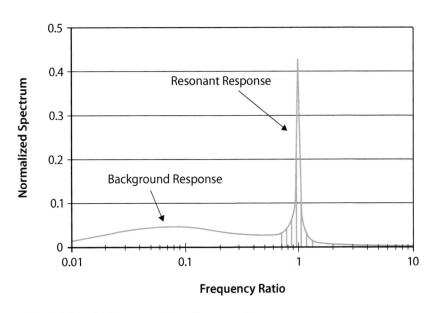

▲ Figure 2.2: Normalized response spectrum of base moment.

cause it to move in its natural modes of vibration. Once the building moves, the acceleration of its mass, which reaches a peak at the extremity of the motion, results in inertial forces on the structure as a consequence of Newton's Second Law (force equals mass times acceleration). The inertial forces due to excitation of each mode of vibration are perfectly correlated over the height of the building and are a function of the natural frequency of the mode, the mass distribution, and the modal deflection shape. Although the inertial loads have as their origin the background loads, there is little correlation on an instant-by-instant basis between the two. On very tall, slender buildings it is usually the inertial loads that dominate over the background loads.

The effect of the resonant response may be seen in the power spectrum of base moment, which is illustrated in Figure 2.2. The background excitation is manifested by a broad range of frequencies appearing in the power spectrum of base moment, whereas the resonant response is concentrated over a narrow band at the natural frequency of the building.

2.5 Serviceability Accelerations

The resonant component of response is also responsible for wind-induced building acceleration. Once the building motion becomes large enough, it becomes perceptible to the occupants. The perception may result from kinaesthetic effects (feeling the motion), auditory cues, or visual cues. Most commonly, the visual cues originate from inside the building (e.g., swinging blinds) and the visual cues are often the first cues to motion. Occupant response to building motion is highly variable and subjective, and can be dependent on a number of factors, such as education about building motion

and past experience of motion, motion sickness susceptibility, and motivation for complaint. There are a number of published guidelines to assist building designers in assessing the acceptability of wind-induced building acceleration, e.g., ISO6897: 1984, ISO10137: 2007, AIJ (1991) and AIJ (2004). Recent trends have been towards shorter return periods for assessment (typically one year) and adjusting the acceptable acceleration based on the natural frequency of vibration of the building. At the time of publication (2013), a state-of-the-art review of occupant response to wind-induced building motion is in the final editing stages for publication by ASCE and CTBUH.

2.6 Vortex Excitation

An important phenomenon for tall buildings is vortex excitation, caused by alternate shedding of vortices from the two sides of the building. As already indicated, these vortices cause the crosswind background loading to become highly correlated up the building's height. The expression of Strouhal gives the frequency f at which the vortices are shed from the side of the building, causing oscillatory crosswind forces at this frequency.

$$f = S\frac{U}{b} \qquad (1)$$

where S = Strouhal number
U = wind speed
b = building width

The Strouhal number is a shape-specific constant with a value typically in the range 0.1 to 0.3. For a square cross-section, it is around 0.10 to 0.14 (depending on height-to-width ratio), and for a rough circular cylinder it is about 0.20. When f matches one of the natural frequencies f_r of the building, resonance occurs, which results in

amplified crosswind response. From Equation 1, this will happen when the wind speed is given by:

$$U = \frac{f_r b}{S} \qquad (2)$$

Figure 2.3 illustrates qualitatively the effect of vortex excitation on the crosswind response of a building. In the absence of vortex excitation the response increases in proportion to wind velocity, raised roughly to the power of two or slightly higher. However, if vortex excitation is present, it results in an amplified response at a speed given by Equation 2. Slightly below this speed, the loads can increase as velocity to a power considerably higher than two. Exponents in the range of three to four are common. The loads reach a peak at the critical speed, and at higher speeds may actually reduce. On very tall buildings, because of their inherently low natural frequencies, it is often impossible to avoid the peak occurring within the design range of wind speeds. Adding stiffness to push the peak further to the right in Figure 2.3, and thus to a velocity safely above the design range, may be prohibitively expensive. In this case it is better to explore ways of reducing the vortex shedding strength via changing shape, or by suppressing the response by using supplementary damping devices.

While the initiation of vortex excitation occurs at the speed given by Equation 1, once the building motions build to sufficient amplitude, the vortex shedding tends to become locked to the frequency of the building motion. As a result, even if the wind speed subsequently changes, the vortex frequency stays locked at the frequency of the building, causing the motions to persist over a range of wind speeds and creating a further amplification of the response.

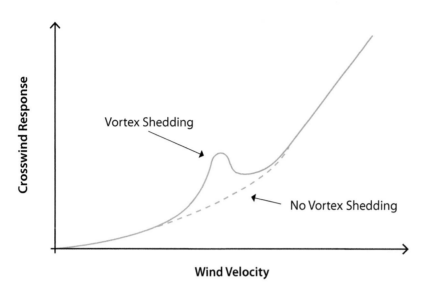

▲ Figure 2.3: Influence of vortex shedding on crosswind response.

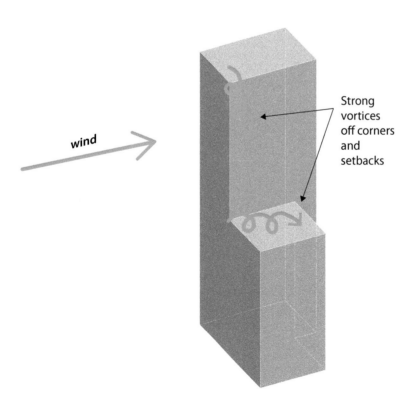

▲ Figure 2.4: Vortices off corners and setbacks create very high local suctions on adjacent surfaces.

2.7 Aerodynamic Damping and Galloping

In principle, the wind forces on a building depend not only on its shape and the characteristics of the ambient wind flow, but also the building's acceleration, velocity, and deflection. For the normal mass densities of tall buildings, the dependencies of wind forces on the tower's acceleration and deflection are very minor and they are usually neglected in wind tunnel studies. However, the dependence on building velocity can become significant in some circumstances; the resulting effect is similar to that of changing the building's damping. As a result it is often called aerodynamic damping.

The aerodynamic damping can be positive, leading to reduced building response, or negative, leading to increased building response. The aerodynamic damping for motions in the along-wind direction is always positive and increases in proportion to mean wind velocity. The aerodynamic damping in the crosswind direction can be of either sign, and when it is negative will amplify the crosswind response. If the negative aerodynamic damping was sufficient to overcome the positive structural damping, then the total damping would be negative and instability would occur.

In this unstable situation, which occurs once a critical wind speed is reached, any small disturbance would be sufficient to trigger growing oscillations of the structure that could reach very damaging amplitudes. This type of instability is termed galloping and should be avoided at all costs. Fortunately, for the typical dimensions and frequencies of most buildings, the critical wind speed for galloping is high enough to be well beyond the design range. However, for slender spires and columns, or for supertall buildings with very low natural frequencies, the potential for the galloping type of dynamic instability should be assessed. Aerodynamic damping effects cannot be measured using the typical aerodynamic model tests, but can only be accurately assessed using aeroelastic test techniques.

2.8 Cladding Loads

Cladding is affected primarily by the exterior local wind pressure acting on a small area of the building envelope, such as the area of a single glazing unit or curtain wall panel. Over such a small area, the exterior wind pressure is highly correlated and can be strongly influenced by local flow phenomena, such as vortices peeling off a building corner, as illustrated in Figure 2.4., which cause high local suctions. Similar vortices cause high local suction at building setbacks, even near the base of a building, and near roof corners and edges, and may persist in the downwind direction and impact other buildings.

Wind tunnel studies to determine cladding loads, therefore, involve models instrumented with many hundreds of pressure taps (sometimes over 1,000), in order to measure the detailed local pressure patterns, both positive (i.e., acting into the building) and negative (i.e., acting in an outward sense and usually referred to as "suctions").

Since cladding responds to the net difference in pressure between its inner and outer surfaces, the internal as well as the external pressure needs to be provided for in design. The internal pressure is a function of leakage paths through the building envelope and the exterior pressures at the points of leakage, as well as any additional pressures created by the building's HVAC systems and stack effect. For buildings with significant openings, such as operable windows left open or windows broken by flying debris created by a

wind storm, internal pressures will tend to be magnified relative to the condition where all windows or other potential openings are closed. It is normal for wind tunnel laboratories to include appropriate allowances for internal pressure effects in their recommended design cladding loads.

2.9 Reynolds Number Effects

An important aerodynamic parameter is the Reynolds number R_e defined by:

$$R_e \equiv \frac{Ub}{\nu} \qquad (3)$$

where U = wind velocity
b = representative building width
ν = kinematic viscosity of air

In principle, the flow patterns around a building, and thus the wind loads on it, are a function of R_e. Therefore, wind tunnel tests run on a small scale model would ideally be run at the same Reynolds number as would be experienced by the full scale building, thus satisfying Reynolds number similarity. In practice, this is completely unfeasible, even in the largest, highest-speed, and most costly wind tunnels.

Fortunately for sharp-edged buildings, which constitute the majority, the flow patterns are dictated by flow separation off the sharp corners and are insensitive to Reynolds number over a very wide range of values. Therefore, for most buildings the necessary relaxation of Reynolds number similarity in wind tunnel tests has little impact on the validity of the results. However, on buildings with curved faces, the flow separation points are less well-defined and can become a function of Reynolds number. In these cases special measures may be taken in an attempt to correct

for the lack of Reynolds number similarity, either by techniques in which the model surface is roughened to artificially achieve a higher "effective Reynolds number," or by making approximate analytical corrections to the wind tunnel results. These techniques are far from exact, especially in the case of cross-wind responses. As a result, for major signature towers, it is not uncommon to supplement the main test program with supplementary tests at the highest achievable Reynolds number, in order to assess the extent to which Reynolds number may be affecting the results.

> For the normal mass densities of tall buildings, the dependences of wind forces on the tower's acceleration and deflection are very minor and they are usually neglected in wind tunnel studies. However, the dependence on building velocity can become significant in some circumstances; the resulting effect is similar to that of changing the building's damping.

3.0 Wind Tunnel Testing Methods

3.1 Simulation of the Natural Wind at Small Scale

In view of the way the mean velocity profile and turbulence characteristics of the wind affect wind loads, it is important that flow in the wind tunnel replicates these factors at model scale. The methods for doing this are well established and typically involve a combination of special flow devices, such as spires at the start of the working section, followed by a length of roughness on the wind tunnel floor, representing the terrain roughness over which the wind flows at full scale. These methods are described in the detailed standards and manuals of practice on wind tunnel testing referenced in the Introduction. Figures 3.1–3.3 show examples of typical test set-ups and Figures 3.4 & 3.5 show typical mean velocity and turbulence intensity profiles generated in the wind tunnel for open terrain compared with target profiles derived from full-scale data. The mean velocity profile shown is the ratio of mean velocity U to the mean velocity at a reference height, in this case 400 meters. In addition to mean velocity and turbulence intensity

> The wind forces on a building model can be affected by the size of the model relative to the cross-sectional area of the working section of the wind tunnel, this being called the blockage effect.

profiles, the power spectrum of turbulence is also simulated, which effectively ensures that the sizes of the turbulent eddies impacting the model have been scaled down to the appropriate scale. The simulation of the wind profiles can be adjusted to various types of terrain roughness, and for most sites it is necessary to change the wind simulation several times during the tests to reflect the variation of upwind conditions with wind direction. It should be noted that most wind tunnel laboratories have a set of several standard wind profiles and pick the closest to that needed for each wind direction. Analytical methods can then be used to correct for minor residual differences between actual and target profiles (see for example Irwin et al. 2005).

The wind forces on a building model can be affected by the size of the model relative to the cross-sectional area of the working section of the wind tunnel. This is called the blockage effect. The airflow is constrained to flow through effectively a smaller cross-sectional area as it flows around the model, and so accelerates, causing the wind forces to be higher than would otherwise be the case.

There are other secondary effects on the wind profiles approaching the model, since they are now subject to pressure gradients. Much of the blockage effect can be eliminated if the reference velocity is measured in the accelerated flow, i.e., directly above the model, but there are also other methods for making blockage corrections to results. In general it is advisable to keep the blockage area to less than 10% of the working cross-sectional area. Open jet wind tunnels can be subject to negative blockage, i.e., the flow effectively decelerates at the model station, and some wind tunnels are made blockage-tolerant by building working section surfaces that are not completely solid.

3.2 Test Methods to Determine Wind Loads on the Structural System

There are several different methods of using the wind tunnel to predict the overall structural loads and responses of tall buildings. The three most common techniques are the high-frequency-balance (HFB) method, the high-frequency-pressure-integration (HFPI) method, and the aeroelastic model method.

The HFB method is also known as the high-frequency force balance method and the high-frequency base balance method. The balance is usually a strain gauge or piezoelectric device, mounted at or near the base of a rigid building model (see Figure 3.6). The balance is capable of measuring instantaneous base moments and, in some balances, base shears. The base moments in particular are closely related to the aerodynamic modal forces acting on the building's lowest modes of vibration. The measurements can be used to determine not only the integrated mean forces on the building, but also the instantaneous fluctuating aerodynamic forces and the modal forces exciting each mode of vibration. These applied aerodynamic forces are combined with the predicted structural dynamic properties of the building to determine the anticipated wind-induced loads and building responses.

The reason the method is called "high-frequency" is because the natural frequency of the model/balance system should be significantly higher than the scaled-prototype natural frequencies of vibration of the building being tested. This ensures that the measured signal is not contaminated by resonant effects from the model/balance system itself.

High-frequency balances come in a number of forms, but all should be able to measure the base bending moments

▲ Figure 3.1: Shanghai Tower, Shanghai, China. Wind tunnel test performed in March 2009. © RWDI

▲ Figure 3.2: Haeundae Beach Towers, Busan, Korea. Wind tunnel testing performed in February 2012. © BMT Fluid Mechanics

> The high-frequency balance method is particularly useful as a tool to explore the effects of changes in the building shape, with a view to optimizing its aerodynamic performance.

▲ Figure 3.3: Empire Tower, Reem Island, Abu Dhabi (future surrounds case). Wind tunnel testing performed in March 2009. © Windtech Consultants Pty. Ltd.

and base torque. Assumptions implicit in the HFB measurement technique are that only the first mode of vibration in each principal direction is important, and that the modal deflection shape increases linearly with height. The first approximation is often acceptable, except for the most slender of buildings, and correction methods exist to minimize errors when the modal deflections depart from the linear shape. An advantage of the high-frequency balance method is that the models are relatively quick and economical to build, and the testing is also relatively quick. Therefore, it is particularly useful as a tool to explore the effects of changes in the building shape, with a view to optimizing its aerodynamic performance.

The HFPI technique became feasible in the 1990s with the development of high-speed electronic pressure scanning systems capable of measuring pressures in the wind tunnel at several hundred locations simultaneously at sample rates of several hundred hertz. With this technique the instantaneous pressure patterns measured on the model are

integrated numerically to obtain the overall mean forces, background forces, and modal forces acting on each mode of vibration. Typically, the same model that was constructed for the cladding pressure test (see Figure 3.7) is used. Each pressure tap on the model is assigned a tributary area. By compiling the simultaneous time histories of pressures, the overall aerodynamic loads on the building are computed. From there, the analysis is the same as the high-frequency balance technique.

An advantage of the HFPI method is that it resolves the variation of aerodynamic force and torque with height, with more precision than the HFB method. This allows the more accurate determination of the building's torsional response, and also enables the response of higher modes of vibration, i.e., beyond the fundamental mode, to be predicted. A limitation of the HFPI method is that for buildings with very complex shapes, or with many fine-scale features, such as lattices, it is not possible to install enough pressure taps in all the required locations to accurately resolve the

overall forces on the building. In these cases, the HFB method is still preferable.

Recent architectural design has seen an increasing number of linked towers or buildings with multiple primary lateral load resisting systems. For this type of structure, it can be necessary to gain a fuller understanding of the distribution of loads acting over the structure(s). In these cases, HFPI can be used if the architecture is simple enough. Otherwise, multiple high-frequency balances can be used with measurements made simultaneously.

3.3 Aeroelastic Model Testing

Both the HFB and HFPI methods use rigid models. This means that they do not directly measure motion-induced forces. Any effects of the building motion through the air, called "aeroelastic effects," are not accounted for with rigid models. A flexible, i.e., aeroelastic, model that simulates the building motions is needed to include the aeroelastic forces. The most important aeroelastic effect

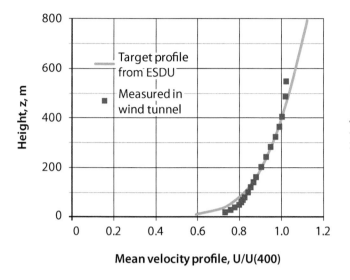

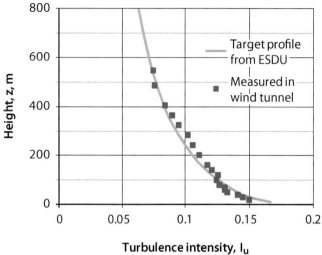

▲ Figure 3.4: Example of mean velocity profile simulation in the wind tunnel.

▲ Figure 3.5: Example of simulation of turbulence intensity in the wind tunnel.

▲ Figure 3.6: High-frequency balance model (left) and example of strain gauge balance (right). © RWDI

▲ Figure3.7: Pressure study model instrumented with pressure taps and tubing. © RWDI

▲ Figure 3.8: Example of aeroelastic model illustrating inner spine for providing stiffness scaling, and segmented outer shell to provide shape and mass scaling. © RWDI

is aerodynamic damping, caused by aerodynamic forces that are proportional to the building's velocity. Aerodynamic damping is normally positive, i.e., it adds to the structural damping already inherent in the building, and therefore tends to reduce building movements. However, there are phenomena such as vortex shedding, galloping, and flutter that can cause the aerodynamic damping to be negative, which causes building motions to be amplified. When either positive or negative aerodynamic damping are deemed to be potentially significant, aeroelastic testing may be conducted to measure the effects directly. An aeroelastic model is one that has not only the correct shape but is also flexible in a way that duplicates, at model scale, the lowest modes of vibration of the real building. Thus, the stiffness, mass distribution, and damping of the model are scaled-down versions of the full-scale stiffness, mass distribution, and damping. Full aeroelastic models are considerably more complex to build than HFB or HFPI models.

For a full aeroelastic simulation the stiffness is modeled by an internal frame or spine, to which the outer shell is attached in segments (see Figure 3.8). The model simulates not only the lowest modes of vibration in each direction and in torsion, but also higher modes. However, for buildings that are relatively stiff in torsion the simulation of torsion can be omitted, which results in simpler model construction. If higher modes of vibration are also unimportant, then the model can be further simplified to the extent that the only point of flexibility is at the base. In this last case a single flexure at the base or a gimbal mounting and spring arrangement may be used. Most commonly, aeroelastic testing is only conducted for tall buildings once the fundamental aerodynamic characteristics have been determined by either HFB or HFPI testing, and the potential for negative aerodynamic damping has been identified.

3.4 Test Methods to Determine Cladding Loads

For the design of the building envelope, predictions of local external pressures exerted by the wind and internal pressure are needed. The model used is rigid, and it is instrumented with up to many hundreds of pressure taps (see Figure 3.7), connected via tubing to electronic pressure transducers. Typically, a higher density of pressure taps will be used for the determination of design cladding pressures compared with an HFPI study, so as to pick up local hot spots of high suction. While the wind tunnel test will measure external pressures, the design cladding pressures should be net pressures that take account of internal pressures in the building. The internal pressures may be determined from code estimates based on assumed leakage scenarios, or from integration of external pressures around the parts of a building where there may be significant leakage. The code approach is more common for nominally sealed buildings, but the use of external pressures is more common when specific scenarios of openings in external façades are being considered. Examples where this might be important include tall residential buildings with accessible balconies, where the external pressures may be transferred to the inside of the building and result in significant pressure differentials across internal walls and partitions.

These pressure differences may affect door operability on a regular basis or, in extreme cases, lead to failures of dividing walls as the pressures are higher than the loads for which an internal wall would normally be designed. The design team for the building needs to make the decision on the likelihood of such an opening occurring during a severe wind storm. In climates where the design wind speeds are dominated by thunderstorms, it is quite likely that a window may be left open, since warning times are much shorter for these storms. If there is an area lower in the building susceptible to impact damage by flying debris, then it may be wise to consider the possibility of an opening in this area. However, in areas where there is a mixed climate, and where the peak wind speeds come from very rare, but forecastable events, such as hurricanes, then it is less likely that there will be openings, except as the result of impact, as building management and/or occupants should be able to ensure that potential building openings are closed.

3.5 Other Types of Wind Related Studies

A wind tunnel model is often used to study not only wind loading but also other wind-related design issues. These will not be described in detail in this guide but are summarized below.

▸ **Building appurtenances.** Spires, architectural lattices, antennae, shading devices, etc. may be prone to wind-induced vibrations. These appurtenances are typically too small to be modeled at the same scale as the building itself. But the wind pressures acting on them, and their susceptibility to vibrations, should be assessed by the wind tunnel consultant and, if necessary, special-purpose wind tunnel tests should be undertaken using a larger-scale model of the appurtenance in question. When large-scale models are used, the requirements to model the full planetary boundary layer in the wind tunnel are necessarily relaxed. The data then require special interpretation to account for the effects of large-scale turbulence that may be missing from the simulation.

▸ **Pedestrian-level winds.** (see ASCE 2003). In these studies, the model is instrumented with wind speed sensors to measure mean and gust speeds around the project at ground level, podium level, and on balconies and terraces. The results, when combined with wind statistics, allow the wind conditions around the project to be compared with criteria for human safety and comfort. Mitigation measures such as massing changes, wind screens, and landscaping can also be evaluated.

▸ **Exhaust dispersion and HVAC system optimization.** The dispersion of building exhausts is frequently examined in the wind tunnel using smoke visualization and tracer gas techniques. The effects of the complex wind patterns on the dispersion are accurately simulated, allowing adverse re-ingestion of exhausts to be avoided. This type of study can also be used to assess the dispersion of odors from kitchens and other areas. The results of pressure studies are also frequently used to provide information on wind pressures at building intakes and exhausts.

▸ **Natural ventilation.** In cases where it is intended to use natural ventilation, the wind pressures that drive the ventilation are critical to the feasibility and effectiveness of the system. Useful information on these pressures is obtained from the wind pressure model. The measured pressures can then be used as boundary conditions for unsteady CFD modeling of the internal flows.

▸ **Roof pavers and gravel.** The lift off of roof pavers or scouring of roof gravel ballast can also be studied using wind tunnel models.

▸ **Icing and Sliding Snow/Ice.** Tall buildings in cold climates can collect snow and ice from snow fall, freezing rain, and in-cloud icing. The collection patterns depend on wind among other factors. Snow or ice sheets may slide off, or be picked up by wind, and cause hazardous conditions at lower levels. Computational methods and physical tests in wind tunnels capable of simulating the snow and ice accumulations on building components are used to investigate these issues.

▸ **Snow or Sand loads.** In cases of entrance roofs, canopies, or buildings with large flat roofs, the irregular distribution of snow loads (in moderate or cold climates) or the irregular accumulation of sand loads (in hot climates near deserts) are influenced by the air flow around tall buildings. Wind tunnel models can be used to learn more about these load distributions.

4.0 Prediction of Load Effects for Strength Design and Serviceability

4.1 Structural Properties of the Building

For the prediction of a building's dynamic response at various wind speeds and directions, it is necessary to know the natural modes of vibration of the structure, including the natural frequencies and modal deflection shapes. It is also necessary to know the damping in each mode. The natural frequencies and mode shapes may be computed using a number of available commercial finite element software packages. However, the results will be influenced by the assumptions made when developing the finite element model.

For the purpose of determining the dynamic response to wind loading, it is important to be aware that, in general, lower frequencies will lead to increased response, which is the opposite situation to that of seismic response. Also, it has been noted that due to the cracking behavior of concrete structures, particularly in elements like link beams, there is a tendency for the stiffness, and thus the frequency, to decrease as the amplitude of motion increases.

In steel buildings, a somewhat similar behavior has been observed, which has been attributed to slippage occurring at connections as amplitudes increase. These changes will clearly affect the response to wind. Therefore it is advisable to assess the sensitivity of wind tunnel predictions to the effect of different cracking or stiffness assumptions. Greater reductions in stiffness may be expected at deflections corresponding to ultimate load conditions than those at which accelerations are assessed.

The damping inherent in the structural systems of tall buildings cannot currently be predicted using detailed analytical methods. Common practice over several decades has been to assume damping ratios of approximately 0.010 to 0.015 for slender steel buildings and about 0.010 to 0.020 for slender concrete buildings. The lower ends of these ranges are applied when assessing building motions; upper end is applied when determining wind loading for structural integrity. Higher values have also been adopted in some cases where ultimate limit state wind speeds have been applied directly without load factors. More detailed empirical relationships have been developed in Japan based on extensive monitoring studies on buildings up to 200 meters in height (Tamura 2012). These show an initial increase of damping ratio as deflections increase, but the damping then levels off beyond a "critical" deflection somewhere in the range of:

$$2 \times 10^{-5} < x_H / H < 1 \times 10^{-4} \qquad (4)$$

where x_H = deflection at the top of the building
H = height of building

Representative values found for the damping ratio were 0.0115 for office buildings of average height 113 meters and 0.0145 for hotels and apartment buildings of average height 100 meters. Other recent data (Willford et al. 2008) indicate that for tall and slender towers above 250 meters in height, where cantilever action dominates over moment frame action, the damping could be below the ranges described above.

As with assumptions concerning stiffness, it is advisable to undertake sensitivity studies of the building response with different damping assumptions. Where lower bound damping assumptions create or accentuate motion or loading problems, the use of supplementary damping systems can be considered as a means of increasing the damping and reducing uncertainty as to what the total damping will be.

4.2 Load Effects

In structural design the important variables for the designer are typically load effects, such as the base bending moments, base shears, base torsion, and corresponding force and torque distributions with height. These global load effects are selected because they are closely correlated with the load levels reached in individual structural members and components. For the design of cladding, the local, peak positive and negative pressures on small areas, such as the area of a cladding element are needed.

For strength design, the load effects may be needed for return periods on the order of 50 to 100 years at service limit state, or 500 to 2,000 years at ultimate limit state. Other load effects, such as accelerations or rotational velocities, may be needed for shorter return periods in the range from a few months up to 10 years in order to evaluate the comfort of occupants. Wind tunnel tests enable a selected wind load effect E to be determined in detail as a function of wind speed and wind direction, but to convert this information into the load effect E_T for the T-year return period requires that the wind tunnel data be combined with the statistics of wind speed and direction at the site. This latter information is usually obtained by extrapolation from the records available from meteorological stations in the area, often from nearby airports. In areas affected by hurricanes and typhoons, the necessary statistics are generated by Monte Carlo simulation.

There are several methods used to combine the wind tunnel data with the meteorological data, and it is important to be aware of the differences. It is also important to realize that the event that causes the T-year wind speed is not necessarily the same event that causes the T-year load effect. Because the wind loading and response of tall buildings is highly dependent on wind direction as well as speed, there is no simple generally applicable explicit relationship between the overall T-year speed and the T-year load effect. To obtain the T-year load effect, special analysis methods are needed for combining wind tunnel data with meteorological data, as explained in the following sections.

4.3 Non-Directional Method

The Non-Directional method is the simplest of the analysis methods, because it simply assumes that the T-year wind speed occurs from all directions, one of which will be the aerodynamically worst direction for the building. Clearly, there is inherent conservatism in this assumption.

Denoting the T-year return-period speed for the building site, including winds from all directions, as U_T then in the Non-Directional method the T-year return period load effect E_T is evaluated as follows:

From the wind tunnel data, which must cover sufficient wind directions to resolve all peak responses (typically 10-degree intervals are sufficient), the load effect $E(U_T)$ at speed U_T is determined for every tested direction. Figure 4.1 illustrates this for the case where E is the base moment, and shows the mean value and the peak positive and negative values for each wind direction evaluated at speed U_T. Note that the overall T-year return period wind speed U_T is taken as constant, independent of wind direction.

The T-year return-period load effect is then taken to be the maximum value of $E(U_T)$ out of all the directions, as illustrated in Figure 4.1 for both negative and positive moments. Provided the

load effect's dependence on speed is monotonic for every direction, i.e., there is no peak response at speeds lower than U_T, the load effect determined by the Non-Directional method will be an upper bound for the T-year return period value E_T. Use of upper bound loads, rather than the true value of E_T would usually result in significant increases in construction cost. For this reason more accurate methods than the simple Non-Directional method are typically used for tall buildings, except where the available meteorological data are insufficient to establish the wind's directional behavior with confidence.

▸ *Advantages:* Very simple. Easily compatible with code-mandated design wind speeds.

▸ *Disadvantages:* Can be very conservative.

4.4 Sector Velocity Method

The Sector Velocity method improves on the Non-Directional method by applying

a different design velocity to each sector of wind direction, using higher wind velocities for directions of higher probability for strong winds (Holmes 1990). The load effect for each wind direction is calculated using these different sector velocities and the highest load effect out of all the directions is taken to be the T-year return period value E_T.

An advantage of this straightforward approach is that it seems intuitive to use higher design speeds for directions where historical records show greater frequency of strong winds, and lower speeds in other sectors. A disadvantage is that there is a theoretical difficulty in selecting appropriate speeds for the various directional sectors. This is because in the general case the overall probability of a given load effect being exceeded is the sum of the probabilities from all the sectors. One does not know in advance, for any particular project, how many sectors will contribute significantly.

The number of sectors that contribute affects the level of probability at which to select the wind speed for each

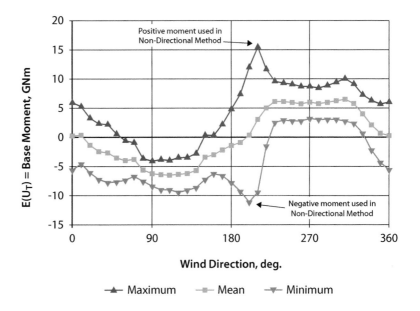

▲ Figure 4.1: Example of Non-Directional method applied to base moment. Note that moments are evaluated at the same wind velocity U_T for all wind directions.

direction, but the number of sectors contributing depends in turn on both the aerodynamics of the building under study and the local wind statistics. Therefore there is no universally "correct" set of sector velocities for all projects at a given location, and engineering judgment plays an important role in selecting the sector velocities for a given project. Nonetheless, provided these complexities are understood, it is still possible to derive a set of sector velocities that give a significantly more accurate and cost-effective result for E_T than the Non-Directional method.

▸ *Advantages:* Simple to understand. Consistent with code-mandated directional wind speeds. Allows relatively straightforward checking of wind tunnel results when directional wind speeds have been specified.

▸ *Disadvantages:* Can rely strongly on engineering judgment and expertise of practitioners to produce the most accurate results.

4.5 Extreme Load Effect Method

The focus of the Non-Directional and Sector Velocity methods is on extreme wind speeds, while what are needed

The Extreme Load Effect method goes directly to the extreme loads without approximation, but it does involve significantly more computation.

in the end for design are extreme load effects. The Extreme Load Effect method, also often referred to as the Storm Passage Method, goes directly to the extreme loads without approximation and in this regard is superior to the two preceding methods, but it does involve significantly more computation. The analysis of meteorological data, or the output of Monte Carlo simulations, allows a complete history of hour-by-hour wind speeds and directions to be generated for the site.

In the Extreme Load Effect method, the wind tunnel data are combined with the meteorological data to produce a corresponding complete hourly history of peak load effects. The peak load effects for each year, month, or for each independent storm are then analyzed using standard extreme value analysis methods, in order to compute directly the load effects as a function of return period (Isyumov et al. 2002). This method is particularly suitable when a Monte Carlo simulation has been used to generate the hour-by-hour site wind speed and direction, because such data are typically well behaved. Wind data derived directly from a meteorological station may contain anomalies due to reading errors, missing records, etc. that will translate directly into anomalies in the extreme load effects. Therefore, in these cases it is important to scan such data first to remove such anomalies or to use smoothed statistical distributions such as is done in the Upcrossing Method described next.

▸ *Advantages:* Very direct method of calculation. Conceptually easy to understand.

▸ *Disadvantages:* Accuracy entirely reliant on quality of input wind data, which limits its use in many areas. Computationally intensive.

4.6 Upcrossing Methods

Analytical methods have also been developed that mathematically mimic the process of the Extreme Load Effect method. These are called Upcrossing methods (Davenport 1977; Lepage and Irwin 1985; Irwin et al. 2005). They use mathematical models of the historical wind statistics, rather than the detailed hour-by-hour history of wind speed and direction, and focus on determining the rate at which a given level of load effect is exceeded.

Upcrossing methods were developed before the Extreme Load Effect method, when less computing power was available. They are still useful, even now in the days of copious computing power, in that where the quality of the meteorological data is not high, the mathematical model of the wind statistics enables anomalies in the raw wind data to be removed or smoothed out before combination with wind tunnel data. Upcrossing methods also have advantages, where data from a number of local stations are combined into a single statistical model of the wind climate to improve statistical reliability. It is important that the mathematical model of the meteorological data used for the Upcrossing method be consistent with extreme value analysis of the wind speeds. In other words, considering the winds from all directions, the model should provide the same wind speed for any given return period as the extreme value analysis of wind speeds.

▸ *Advantages:* Combines the advantages of the Extreme Load Effect approach with a statistical fit to wind climate data.

▸ *Disadvantages:* Can be difficult to explain to statutory authorities. Not easy for third parties to check results.

5.0 Format for Comparing Wind Tunnel Results

5.1 Types of Comparison

For major projects, it is quite common for wind tunnel tests to be undertaken at two independent laboratories as a means of improving the reliability of the end results. Therefore it is advantageous if the laboratories involved can present their results in similar formats to allow for quick comparisons to be made. The formats should facilitate five types of comparison, each corresponding to links in the Alan G. Davenport Chain:

(i) Comparison of wind climate models.
(ii) Comparison of wind velocity and turbulence profiles in the flow approaching the wind tunnel model.
(iii) Comparison of the purely aerodynamic data, independent of any wind climate statistics, or building dynamic properties.
(iv) Comparison of wind-induced loads and responses after incorporation of building dynamic properties.
(v) Comparison of final predicted loads and responses at full scale for selected return periods.

Breaking the comparisons down in this way makes it easier to identify sources for any differences that may arise.

5.2 Wind Climate Models, Velocity Profiles, and Turbulence

Wind climate models are most easily compared if the following data are presented:

(i) A plot and/or table of overall predicted mean wind speed, regardless of direction, versus return period at a selected reference height (see Figure 1.5).
(ii) Tables or plots of mean wind speed versus wind direction for selected levels of probability of the speed being exceeded. Typically the number of directions would be in the range of 16 to 36

and the suggested levels of probability of exceedence per hour are 10^{-4}, 10^{-5}, 10^{-6}, and 10^{-7}.
(iii) Tables or plots of probability versus wind direction of selected wind speeds being exceeded. Threshold speeds could be zero and those for return periods of 5, 50, and 500 years as examples (see Figure 1.6).

The mean velocity and turbulence profiles can be compared using formats similar to those of Figures 3.4 & 3.5, supplemented by information on turbulence integral scales at selected heights.

5.3 Aerodynamic Data

It is important to be able to compare the data from the wind tunnel tests themselves, before they are combined with meteorological statistics. The key data for a tall building will be in the form of aerodynamic base moments, base shears, and base torsions. It is good practice to have these presented somewhere in the wind tunnel report, in the form of non-dimensional coefficients as a function of wind direction. Non-dimensional quantities are preferred, since they facilitate direct comparison of results from different laboratories. Barring Reynolds number effects, the values of the non-dimensional coefficients are independent of the wind speed and model scale used for the tests, as long as common reference heights, dimensions, and locations are used. Also, in the absence of Reynolds number effects, they do not change with the transition from model to full scale. The non-dimensional aerodynamic coefficients for base shear are C_{AFx} and C_{AFy} for shear in the x and y directions respectively, and are defined as:

$$C_{AFx} = \frac{F_{Ax}}{\frac{1}{2}\rho U^2 Hb} \qquad (5)$$

$$C_{AFy} = \frac{F_{Ay}}{\frac{1}{2}\rho U^2 Hb} \qquad (6)$$

where F_{Ax} = aerodynamic base shear force in the x direction on the model
F_{Ay} = aerodynamic base shear force in the y direction on the model
ρ = air density in the wind tunnel test
U = reference mean wind speed in the wind tunnel at a selected reference location, well away from the aerodynamic influence of the study building or any others in the locality
H = height of building model
b = a selected representative width of the building model

Similarly the base moments, as a function of wind direction, can be expressed in the form of the non-dimensional coefficients C_{AMx}, C_{AMy}, and C_{AMz}, defined by:

$$C_{AMx} = \frac{M_{Ax}}{\frac{1}{2}\rho U^2 H^2 b} \qquad (7)$$

$$C_{AMy} = \frac{M_{Ay}}{\frac{1}{2}\rho U^2 H^2 b} \qquad (8)$$

$$C_{AMz} = \frac{M_{Az}}{\frac{1}{2}\rho U^2 Hb^2} \qquad (9)$$

where M_{Ax} = aerodynamic base moment about x axis on the model
M_{Ay} = aerodynamic base moment about y axis on the model
M_{Az} = aerodynamic base moment about the vertical z axis on the model (torsion)

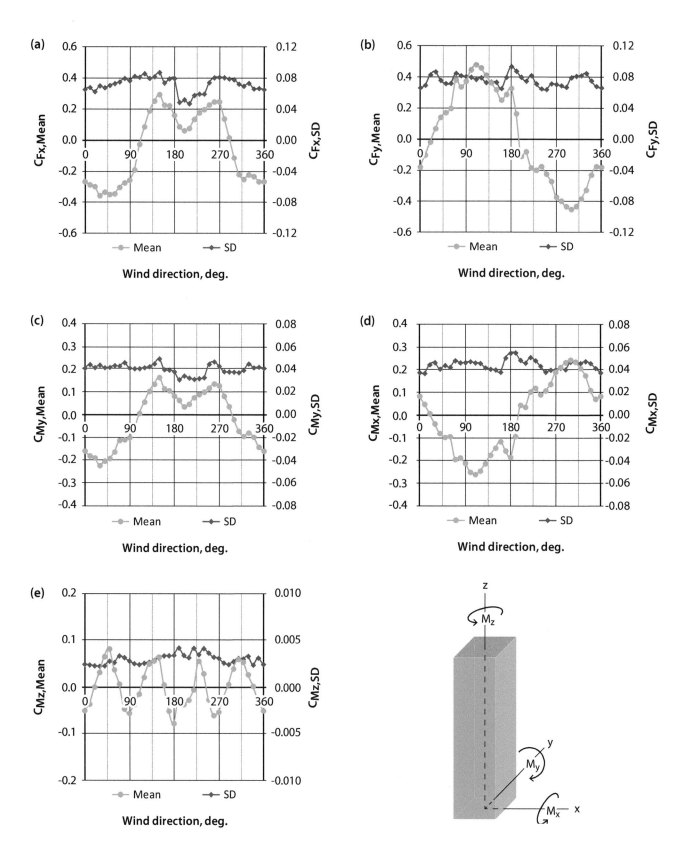

▲ Figure 5.1: Examples of mean and standard deviations of aerodynamic force and moment coefficients as a function of wind direction.

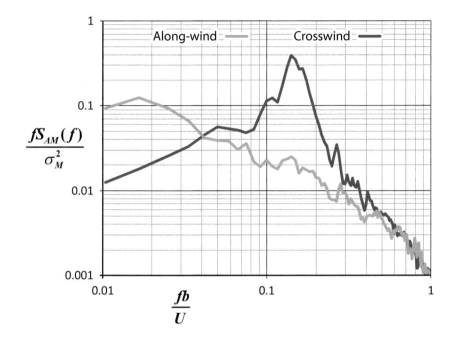

$$\frac{fS_{AM}(f)}{\sigma_M^2}$$

$$\frac{fb}{U}$$

▲ Figure 5.2: Examples of non-dimensional power spectra of base moments due to aerodynamic forces in the along-wind and crosswind directions.

To compare detailed pressure coefficients from two laboratories for every pressure tap is a time-consuming undertaking. However, comparisons for a few selected taps should be sufficient to gauge the general level of agreement.

Mean and standard deviation values of these aerodynamic coefficients, prior to being combined with any dynamic properties of the full-scale building, should be readily comparable between laboratories as a function of wind azimuth. Example plots of the mean and standard deviation of these aerodynamic coefficients as a function of wind direction are shown in Figure 5.1.

Graphs of non-dimensional power spectra, $\overline{S}_{AM}$, of aerodynamic base moments versus non-dimensional frequency, $\overline{f}$, should also be available for critical wind directions. These non-dimensional quantities are defined by

$$\overline{S}_{AM} \equiv \frac{fS_{AM}(f)}{\sigma_M^2} \qquad (10)$$

and

$$\overline{f} \equiv \frac{fb}{U} \qquad (11)$$

where $S_{AM}(f)$ is the power spectrum of aerodynamic base moment and f is the frequency. Examples of non-dimensional spectra of base moments due to wind forces in the along-wind and crosswind directions are shown in Figure 5.2. A similar plot of the power spectrum of base torque can also be presented in non-dimensional form.

For local cladding pressures the data from the wind tunnel should be put into the form of pressure coefficient, C_P, defined as follows:

$$C_P = \frac{p - p_0}{\frac{1}{2}\rho U^2} \qquad (12)$$

In this expression p = local pressure at the pressure tap and p_0 = constant reference static pressure at the reference height. The pressure at the pressure tap fluctuates rapidly in time and typically the mean, standard deviation, and peak

positive and negative values of pressure coefficient are determined as functions of wind direction, usually at every 10 degrees of wind azimuth. Peak values of C_P should be the peak values expected statistically during a period of about one hour at full scale.

Since there are typically many hundreds of pressure taps used in the tests, and since tests are usually undertaken for 36 wind directions, the full data set of pressure coefficients for a building is a substantial body of data. To compare detailed pressure coefficients from two laboratories for every pressure tap is a time-consuming undertaking. However, comparisons for a few selected taps should be sufficient to gauge the general level of agreement. It should be noted that local pressure coefficients can be extremely sensitive to the position of the tap and wind direction, and this needs to be borne in mind when making comparisons of local

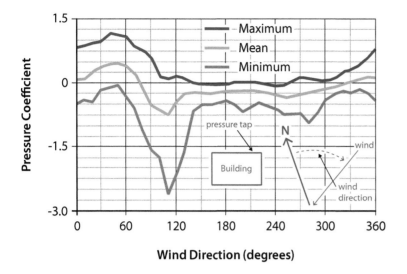

▲ Figure 5.3: Example of the variation of mean, peak maximum, and peak minimum pressure coefficient with wind direction.

pressure coefficients from two different model tests. Figure 5.3 gives examples of the variation of mean and peak negative and positive pressure coefficients with wind direction.

5.4 Predicted Building Response Variations

From the aerodynamic data and the structural properties of the building, the building response may be computed. The structural properties are normally provided to the wind tunnel laboratory by the structural designer. For the purposes of the wind tunnel analysis, they are most conveniently in the form of the natural frequencies and deflected shapes of the lower modes of vibration of the building, as well as the distributions with height of the mass, and polar moments of inertia of mass. In addition, the damping ratio of the lower modes of vibration of the building are needed. With this information there are a number of useful responses of the building that can be predicted as functions of wind speed and direction using random vibration theory. Important overall measures of the response of the building are the

base bending and torsional moments, and the base shears, including the effects of dynamic amplification. Again it facilitates comparison between predictions from different laboratories if the data are expressed in non-dimensional form. The various responses, expressed as non-dimensional coefficients again in order to facilitate comparison between different laboratories, are:

$$C_{Fx} = \frac{F_x}{\frac{1}{2}\rho U^2 Hb} \tag{13}$$

$$C_{Fy} = \frac{F_y}{\frac{1}{2}\rho U^2 Hb} \tag{14}$$

$$C_{Mx} = \frac{M_x}{\frac{1}{2}\rho U^2 H^2 b} \tag{15}$$

$$C_{My} = \frac{M_y}{\frac{1}{2}\rho U^2 H^2 b} \tag{16}$$

$$C_{Mz} = \frac{M_z}{\frac{1}{2}\rho U^2 Hb^2} \tag{17}$$

where the base shears F_x and F_y and base moments M_x, M_y, and M_z now include the effects of dynamic amplification. To facilitate comparison of response predictions from different laboratories, the peak values of these non-dimensional responses should be plotted and/or tabulated versus wind direction for one or more constant reference wind speeds, representative of the service, or design, level speed. The reference speed is the mean speed at the site at a selected reference height. Peak values may be taken as the expected peak value over a period of one hour. Note that since the mean forces and moments are unaffected by dynamic response, the mean values will be the same as the mean values of aerodynamic data in Section 5.3.

It should also be noted that the plots of peak non-dimensional responses just described do not incorporate any information on the directionality of the wind, since the same wind speed is used for all directions. Therefore, provided both laboratories are using the same constant equivalent full-scale speed and the same reference location for the speed, their results should be directly comparable, regardless of what differences there may be in their statistical models of the wind climate. Figure 5.4 gives an example of a comparison of the key base bending moment coefficients C_{Mx} and C_{My} from two different laboratories for a particular tall building project with reference height $H = 500$ meters, reference width $b = 75$ meters and reference site velocity $U = 30$ m/s at a height of 500 meters.

In this project, the data sets from the two wind tunnel laboratories agreed fairly well, but the final predictions of loads for various return periods differed. Because of the vagaries of meteorological data, and the different ways of interpreting such data, this is not uncommon. However, having identified that the basic wind tunnel data were in agreement, it was possible to focus on the question of how to interpret the meteorological data. This interpretation process is critical, since the crosswind response of a tall building may vary with wind velocity to power four or higher. For example, what may appear to be a relatively minor 5% difference in predicted design wind speed can lead to 20% to 25% difference in design base moments.

Another response of importance is the peak acceleration in the upper floors of the building. Acceleration is also extremely sensitive to wind speed, but if the test laboratories present results for the same constant full scale speed for all directions then their results should be readily comparable. Therefore a convenient medium for comparison is a plot of predicted peak acceleration in milli-g versus wind direction for selected wind speeds, such as the 1-year and 10-year speeds. The comparisons may consist of the overall resultant acceleration at an agreed-upon location in the floor plan, or could include more detail in the form of plots of individual sway and torsional components.

5.5 Responses Versus Return Period

The end product desired from wind tunnel testing is the prediction of response versus return period or mean recurrence interval. This involves combining the information on building responses with the statistics of the wind climate for the site as described in Section 4.0. The results of this analysis may best be summarized by plots and tables of base moments, base forces, and peak accelerations at selected levels in the building-versus-return period. Figures 5.5 & 5.6 show examples of base moments and accelerations plotted against return period. It is also useful to provide a summary of the directionality reduction factors, defined as the ratio of the

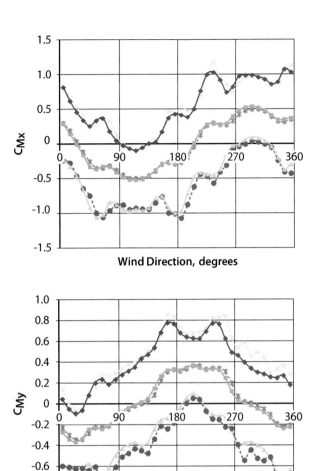

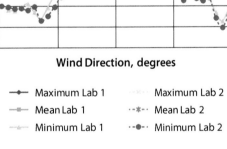

▲ Figure 5.4: Example of a comparison of non-dimensional base bending moment coefficients obtained from two independent wind tunnel laboratories at a selected full-scale reference wind speed of 30 m/s.

predicted responses to those obtained by the Non-Directional method.

5.6 Quality Assurance

Wind tunnel testing involves many steps, including: model construction; wind simulation; undertaking appropriate measurements in the wind tunnel;

recording and analyzing the data using customized software; synthesis of wind tunnel data with full-scale building properties and meteorological statistics; and reporting. It involves significant sized teams of engineers, technologists, and project managers.

Therefore, as with any complex procedure, the issues of coordination

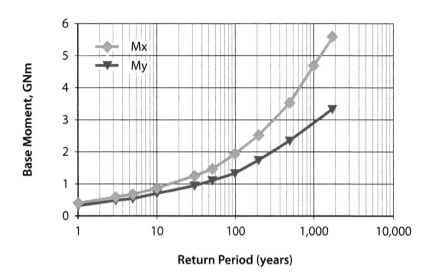

▲ Figure 5.5: Example of variation of base bending moments versus return period for 2% damping ratio for a dynamically sensitive building.

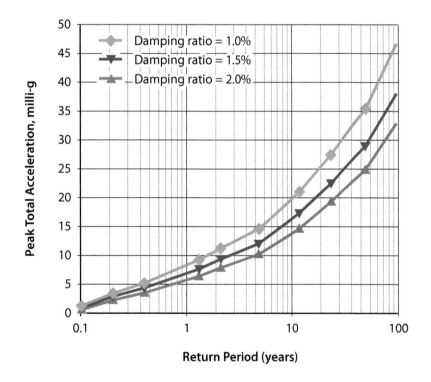

▲ Figure 5.6: Example of variation of peak total acceleration with return period for various damping ratios.

and quality assurance are important. The wind tunnel laboratories undertaking the testing of tall buildings should have appropriate coordination and quality assurance procedures in place. These should be targeted at assuring accuracy and avoiding human error, instrumentation errors, software errors, and communication errors.

General quality assurance measures should include, but not necessarily be limited to the following activities: involvement of an experienced and knowledgeable wind engineer in defining the scope and methodology of the test program; clear assignment of responsibility to other personnel for the various details of the test program; proper record keeping of all testing activities, including photographs of model test set-ups; checks undertaken of all models and analyses by a second independent person specifically assigned for quality assurance purposes; regular maintenance of the wind tunnel facilities to ensure proper flow charac-teristics; periodic checks of wind tunnel repeatability using standard test models; proper maintenance and record keeping of all software used; and a clear path of communication between the wind tunnel laboratory and the design team.

There are many other more detailed aspects to quality assurance, such as the selection of model scale, accuracy of the study model, appropriate modeling of surrounds, simulation of wind profiles, blockage effects, instrumentation performance characteristics, numbers of pressure taps, Reynolds number limitations, etc. For discussions of these, refer to these guidelines and manuals of practice: ASCE, 1999; AWES, 2001; BCJ, 1993 & 2008; KCTBUH, 2009; and ASCE 49–12, 2012.

6.0 Use of Wind Tunnel Results

One of the benefits of wind tunnel tests of tall buildings is that they can assist the designers to arrive at a better optimized structural system through accurate knowledge of the wind loads and the building's response to them. In some cases they may also be used to help optimize the shape.

The decision-making around such optimization studies can be greatly aided by gaining a physical understanding of the factors causing the highest loads or highest accelerations. For buildings where the along-wind response governs the wind loading on the main structural system, the loads from the wind tunnel are quite likely to be similar to, or somewhat less than, those calculated by building code formulae. This is because most building code formulae were developed using conservative assumptions for the along-wind loading case. Along-wind loading tends not to be as sensitive as crosswind to shape, making the value of exploring shape changes less obvious.

However, for very tall and slender towers, the crosswind response is often dominant, and the loads may well exceed those from the building code by a substantial margin. This should not be surprising, because most code formulae simply do not allow for crosswind responses. Since crosswind loading is highly sensitive to shape, a wind tunnel program focusing on shape changes could be very worthwhile. Also, whereas stiffening a building will invariably reduce along-wind loading, it is possible in rare cases for it to have the opposite effect for crosswind responses. This is useful knowledge for the structural designer.

Other factors that will be identified in the wind tunnel tests are the critical wind directions causing peak loading, and wake buffeting or channeling effects caused by nearby buildings.

In some cases buildings have been re-oriented so that the wind directions for highest aerodynamic loading do not coincide with the most common directions for strong winds.

For many tall buildings, one of the most challenging tasks is keeping the building motions within an acceptable range for human comfort. Criteria for what is acceptable for office and residential buildings are described in ISO, 2007, and in that document are based on the 1-year recurrence interval peak acceleration. Other criteria that vary according to the target quality of the building have been published by the Architectural Institute of Japan, AIJ, 1991. The variables at the design team's disposal for controlling motions are shape, mass, stiffness, and damping. Sometimes a combination of all these needs to be brought into play to arrive at an acceptable solution. The HFB method is particularly well suited to undertaking the type of parametric studies needed to explore the effects of these factors.

To reduce building responses to wind, supplementary damping devices are increasingly being used. Damping devices reduce the resonant component of the building response, with the resonant response being approximately inversely proportional to the square root of the damping ratio. Damping devices include very simple liquid sloshing dampers, mechanical tuned mass dampers, computer-controlled active mass dampers, and distributed viscous dampers, among others. The majority of damping systems rely on a few damping devices, and these are thus used only to control building motions. Distributed damping systems, with their inherent redundancy, have been used to provide supplementary damping for the reduction of design loads as well as serviceability accelerations.

For tall buildings with responses very sensitive to wind speed, structural reliability is of the essence. Building codes that use the load and resistance factor approach to establishing design loads generally assume that wind loads vary as wind speed squared. This assumption is implicit in the assumed load factor for wind, which, depending on the code, may lie in the range of 1.4 to 1.6. For a building with responses varying with wind speed to power three or four – which is possible when crosswind response governs – a load factor as high as two or above may be needed to achieve the normal target level of structural reliability. Alternatively, it is becoming an increasingly common practice to evaluate the loads directly at the wind speed corresponding to the ultimate design scenario, in which case the load factor is 1.0.

Supplementary damping devices are being used increasingly to reduce building responses to wind. Damping devices reduce the resonant component of the building response, with the resonant response being approximately inversely proportional to the square root of the damping ratio.

Bibliography

AIJ. (1991) *Guidelines for the Evaluation of Habitability to Building Vibration.* Architectural Institute of Japan.

AIJ. (2004) *Guidelines for the Evaluation of Habitability to Building Vibration.* Architectural Institute of Japan Recommendations, AIJ-GEH-2004: Tokyo.

ASCE. (1999) *Wind Tunnel Studies of Buildings and Structures.* ASCE Manuals and Reports on Engineering Practice, 67. American Society of Civil Engineers: Reston.

ASCE. (2003) *Outdoor Human Comfort and its Assessment; State of the Art Report.* ASCE Task Committee on Outdoor Human Comfort. American Society of Civil Engineers: Reston.

ASCE 49–12. (2012) *Wind Tunnel Testing for Buildings and Other Structures.* American Society of Civil Engineers: Reston.

AWES. (2001) *Quality Assurance Manual on Environment Wind Studies of Buildings.* Australasian Wind Engineering Society. QAM-1-2001. www.awes.org.

BCJ. (1993 & 2008), *A Guide to Wind Tunnel Tests of Buildings for Practictioners.* Building Center of Japan (In Japanese). Chinese version, 2011.

Davenport, A. G. (1977) "The Prediction of Risk under Wind Loading." International Conference of Structural Safety and Reliability. Munich, pp. 511–538.

Davenport, A. G. (1982) "The Interaction of Wind and Structures." *Engineering Meteorology.* Elsevier Scientific Publishing Company: Amsterdam, pp. 557–572.

ESDU. (1993) *Strong Winds in the Atmospheric Boundary Layer, Parts 1 And 2.* Items 82026 and 83045. Issued September 1982 and November 1983 respectively. Engineering Sciences Data Unit, ESDU International: London.

Holmes, J.D. (1990). "Directional Effects on Extreme Wind Loads," *Civil Engineering Transactions,* 32 No. 1. Institution of Engineers: Australia.

Irwin, P. A., Garber, J. J., & Ho, E. (2005) "Integration Of Wind Tunnel Data With Full Scale Wind Climate," *Proceedings of the 10th Americas Conference on Wind Engineering.* Baton Rouge, Louisiana, 31 May – 4 June.

International Organization for Standardization. (1984) *Guidelines for the Evaluation of the Response of Occupants of Fixed Structures, Especially Buildings and Off-shore Structures, to Low-frequency Horizontal Motion (0,063 to 1 Hz) ISO 6897: 1984.* International Organization for Standardization: Geneva.

International Organization for Standardization. (2007). *Bases for Design of Structures – Serviceability of Buildings and Walkways Against Vibrations. ISO 10137: 2007.* International Organization for Standardization: Geneva.

Isyumov, N., Mikitiuk, M. J., Case, P. C., Lythe, G. R., & Welburn, A. (2002) *Predictions of Wind Loads and Responses from Simulated Tropical Storm Passages.* Alan Davenport Symposium, University of Western Ontario, London, Ontario, 19–22 June.

KCTBUH. (2009) *High-Rise Buildings Design Guide for Wind Effects.* Korean Council on Tall Buildings and Urban Habitat, Engineering Dept. 341, Korea University: Seoul.

Lepage, M. F. & Irwin, P. A. (1985) "A Technique for Combining Historical Wind Data with Wind Tunnel Tests to Predict Extreme Wind Loads," *Proceedings of the 5th U.S. National Conference on Wind Engineering.* Lubbock, 6–8 November.

Peterka, J. A. & Shahid, S. (1998) *Design Gust Speeds In The United States.* J. Struct. Engrg., 124(2), pp. 207–214.

Qiu, X., Lepage, L., Sifton, V., Tang, V., and Irwin, P. (2005) *"Extreme Wind Profiles in The Arabian Gulf Region." Proceedings of the 6th Asia Pacific Conference on Wind Engineering.* Seoul, Korea, September.

Tamura, Y. (2012) "Amplitude Dependency of Damping in Buildings and Critical Tip Drift Ratio." *International Journal of High-Rise Buildings,* March, Vol. 1, No. 1, pp. 1–13.

Vickery, P. J., Wadhera, D., Galsworthy, J., Peterka, J. A., Irwin, P. A., & Griffis, L. A. (2010) "Ultimate Wind Load Design Gust Speeds in the United States for Use in ASCE 7." *ASCE Journal of Structural Engineering 136(5),* pp. 613–625.

Willford, M., Whittaker, A., & Klemencic, R. (2008) *Recommendations for the Seismic Design of High-Rise Buildings.* Council on Tall Buildings and Urban Habitat: Chicago, p. 20.

CTBUH Height Criteria

The Council on Tall Buildings and Urban Habitat is the official arbiter of the criteria upon which tall building height is measured, and the title of "The World's (or Country's, or City's) Tallest Building" determined. The Council maintains an extensive set of definitions and criteria for measuring and classifying tall buildings which are the basis for the official "100 Tallest Buildings in the World" list (see pages 42–45).

What is a Tall Building?

There is no absolute definition of what constitutes a "tall building." It is a building that exhibits some element of "tallness" in one or more of the following categories:

▸ **Height relative to context:** It is not just about height, but about the context in which it exists. Thus, whereas a 14-story building may not be considered a tall building in a high-rise city such as Chicago or Hong Kong, in a provincial European city or a suburb this may be distinctly taller than the urban norm.

▸ **Proportion:** Again, a tall building is not just about height, but also about proportion. There are numerous buildings which are not particularly high, but are slender enough to give the appearance of a tall building, especially against low urban backgrounds. Conversely, there are numerous big/large-footprint buildings which are quite tall, but their size/floor area rules them out as being classed of a tall building.

▸ **Tall Building Technologies:** If a building contains technologies which may be attributed as being a product of "tall" (e.g., specific vertical transport technologies, structural wind bracing as a product of height, etc.), then this building can be classed as a tall building.

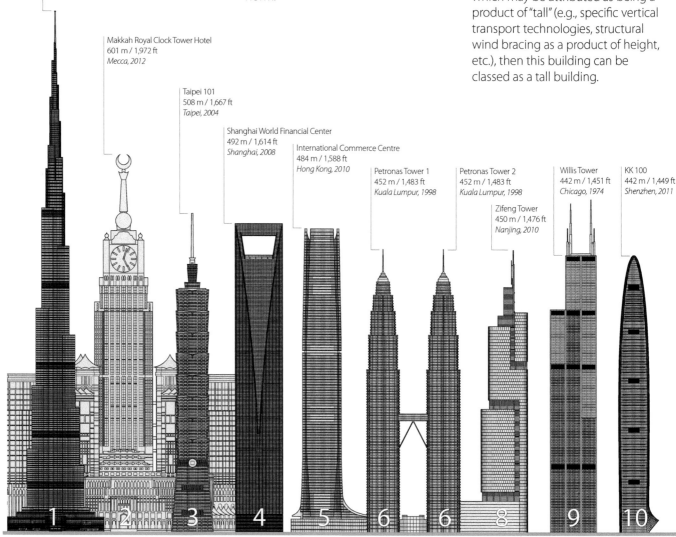

Burj Khalifa
828 m / 2,717 ft
Dubai, 2010

Makkah Royal Clock Tower Hotel
601 m / 1,972 ft
Mecca, 2012

Taipei 101
508 m / 1,667 ft
Taipei, 2004

Shanghai World Financial Center
492 m / 1,614 ft
Shanghai, 2008

International Commerce Centre
484 m / 1,588 ft
Hong Kong, 2010

Petronas Tower 1
452 m / 1,483 ft
Kuala Lumpur, 1998

Petronas Tower 2
452 m / 1,483 ft
Kuala Lumpur, 1998

Zifeng Tower
450 m / 1,476 ft
Nanjing, 2010

Willis Tower
442 m / 1,451 ft
Chicago, 1974

KK 100
442 m / 1,449 ft
Shenzhen, 2011

1   2   3   4   5   6   6   8   9   10

▲ Diagram of the World's Tallest 20 Buildings according to the CTBUH Height Criteria of "Height to Architectural Top" (as of April 2013).

Although number of floors is a poor indicator of defining a tall building due to the changing floor-to-floor height between differing buildings and functions (e.g., office versus residential usage), a building of 14 or more stories – or over 50 meters (165 feet) in height – could perhaps be used as a threshold for considering it a "tall building."

What are Supertall and Megatall Buildings?

The CTBUH defines "supertall" as a building over 300 meters (984 feet) in height, and a "megatall" as a building over 600 meters (1,968 feet) in height. Although great heights are now being achieved with built tall buildings – in excess of 800 meters (2,600 feet) – as of April 2013 there are only approximately 70 supertall and 2 megatall buildings completed and occupied globally.

How is a tall building measured?

The CTBUH recognizes tall building height in three categories:

▸ **Height to Architectural Top:**
Height is measured from the level[1] of the lowest, significant,[2] open-air,[3] pedestrian[4] entrance to the architectural top of the building, including spires, but not including antennae, signage, flagpoles, or other functional-technical equipment.[5] This measurement is the most widely utilized and is employed to define the Council on Tall Buildings and Urban Habitat (CTBUH) rankings of the "World's Tallest Buildings."

▸ **Highest Occupied Floor:**
Height is measured from the level[1] of the lowest, significant,[2] open-air,[3] pedestrian[4] entrance to the finished floor level of the highest occupied[6] floor within the building.

▸ **Height to Tip:**
Height is measured from the level[1] of the lowest, significant,[2] open-air,[3] pedestrian[4] entrance to the highest point of the building, irrespective of material or function of the highest element (i.e., including antennae, flagpoles, signage, and other functional-technical equipment).

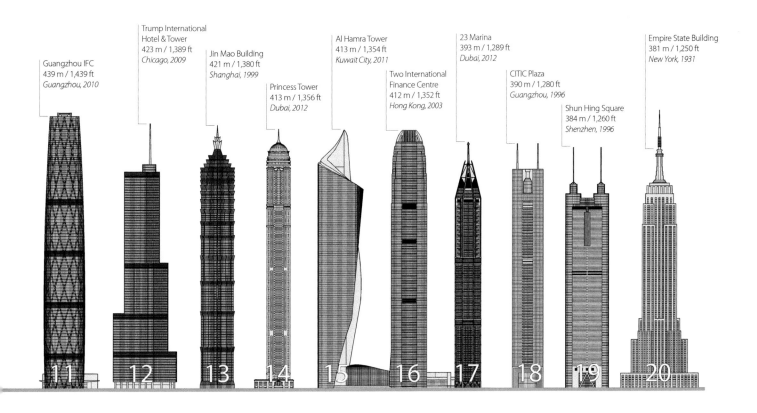

Guangzhou IFC
439 m / 1,439 ft
Guangzhou, 2010

Trump International Hotel & Tower
423 m / 1,389 ft
Chicago, 2009

Jin Mao Building
421 m / 1,380 ft
Shanghai, 1999

Princess Tower
413 m / 1,356 ft
Dubai, 2012

Al Hamra Tower
413 m / 1,354 ft
Kuwait City, 2011

Two International Finance Centre
412 m / 1,352 ft
Hong Kong, 2003

23 Marina
393 m / 1,289 ft
Dubai, 2012

CITIC Plaza
390 m / 1,280 ft
Guangzhou, 1996

Shun Hing Square
384 m / 1,260 ft
Shenzhen, 1996

Empire State Building
381 m / 1,250 ft
New York, 1931

11 12 13 14 15 16 17 18 19 20

Number of Floors:
The number of floors should include the ground floor level and be the number of main floors above ground, including any significant mezzanine floors and major mechanical plant floors. Mechanical mezzanines should not be included if they have a significantly smaller floor area than the major floors below. Similarly, mechanical penthouses or plant rooms protruding above the general roof area should not be counted. Note: CTBUH floor counts may differ from published accounts, as it is common in some regions of the world for certain floor levels not to be included (e.g., the level 4, 14, 24, etc. in Hong Kong).

Building Usage:
What is the difference between a tall building and a telecommunications/observation tower?

▶ A tall "building" can be classed as such (as opposed to a telecommunications/observation tower) and is eligible for the "tallest" lists if at least 50 percent of its height is occupied by usable floor area.

Single-Function and Mixed-Use Buildings:

▶ A single-function tall building is defined as one where 85 percent or more of its total floor area is dedicated to a single usage.

▶ A mixed-use tall building contains two or more functions (or uses), where each of the functions occupy a significant proportion[7] of the tower's total space. Support areas such as car parks and mechanical plant space do not constitute mixed-use functions. Functions are denoted on CTBUH "tallest" lists

in descending order, e.g., "hotel/office" indicates hotel above office function.

Building Status:

▶ **Complete (Completion):**
A building is considered to be "complete" (and added to the CTBUH Tallest Buildings lists) if it fulfills all of the following three criteria: (i) topped out structurally and architecturally, (ii) fully clad, and (iii) open for business, or at least partially occupiable.

▶ **Under Construction (Start of Construction):**
A building is considered to be "under construction" once site clearing has been completed and foundation/piling work has begun.

▶ **Topped Out:**
A building is considered to be "topped out" when it is under construction, and has reached its full height both structurally and architecturally (e.g., including its spires, parapets, etc.).

▶ **Proposed (Proposal):**
A building is considered to be "proposed" (i.e., a real proposal) when it fulfills all of the following criteria: (i) has a specific site with ownership interests within the building development team, (ii) has a full professional design team progressing the design beyond the conceptual stage, (iii) Has obtained, or is in the process of obtaining, formal planning consent/legal permission for construction, and (iv) has a full intention to progress the building to construction and completion.

▶ **Vision:**
A building is considered to be a "vision" when it either: (i) is in the early stages of inception and does not yet fulfill the criteria under the "proposal" category, or (ii) was a proposal that never advanced to the construction stages, or (iii) was a theoretical proposition.

▶ **Demolished:**
A building is considered to be "demolished" after it has been destroyed by controlled end-of-life demolition, fire, natural catastrophe, war, terrorist attack, or through other means intended or unintended.

Structural Material:

▶ A **steel** tall building is defined as one where the main vertical and lateral structural elements and floor systems are constructed from steel.

▶ A **concrete** tall building is defined as one where the main vertical and lateral structural elements and floor systems are constructed from concrete.

▶ A **composite** tall building utilizes a combination of both steel and concrete acting compositely in the main structural elements, thus including a steel building with a concrete core.

▶ A **mixed-structure** tall building is any building that utilizes distinct steel or concrete systems above or below each other. There are two main types of mixed structural systems: a **steel/concrete** tall building indicates a steel structural system located above a concrete structural system, with the opposite true of a **concrete/steel** building

▶ **Additional Notes on Structure:**
(i) If a tall building is of steel construction with a floor system of concrete planks on steel beams, it is considered a **steel** tall building.
(ii) If a tall building is of steel construction with a floor system of a concrete slab on steel beams, it is considered a **steel** tall building.
(iii) If a tall building has steel columns plus a floor system of concrete beams, it is considered a **composite** tall building.

---

[1] Level: finished floor level at threshold of the lowest entrance door.

[2] Significant: the entrance should be predominantly above existing or pre-existing grade and permit access to one or more primary uses in the building via elevators, as opposed to ground floor retail or other uses which solely relate/connect to the immediately adjacent external environment. Thus entrances via below-grade sunken plazas or similar are not generally recognized. Also note that access to car park and/or ancillary/support areas are not considered significant entrances.

[3] Open-air: the entrance must be located directly off of an external space at that level that is open to air.

[4] Pedestrian: refers to common building users or occupants and is intended to exclude service, ancillary, or similar areas.

[5] Functional-technical equipment: this is intended to recognize that functional-technical equipment is subject to removal/addition/change as per prevalent technologies, as is often seen in tall buildings (e.g., antennae, signage, wind turbines, etc. are periodically added, shortened, lengthened, removed, and/or replaced).

[6] Highest occupied floor: this is intended to recognize conditioned space which is designed to be safely and legally occupied by residents, workers, or other building users on a consistent basis. It does not include service or mechanical areas which experience occasional maintenance access, etc.

[7] This "significant proportion" can be judged as 15 percent or greater of either: (i) the total floor area, or (ii) the total building height, in terms of number of floors occupied for the function. However, care should be taken in the case of supertall towers. For example a 20-story hotel function as part of a 150-story tower does not comply with the 15 percent rule, though this would clearly constitute mixed-use.

100 Tallest Buildings in the World (as of April 2013)

The Council maintains the official list of the 100 Tallest Buildings in the World, which are ranked based on the height to architectural top, and includes not only completed buildings, but also buildings currently under construction. However, a building does not receive an official ranking number until it is completed (see criteria, pages 38–41).

Color Key:
Buildings listed in black are completed and officially ranked.
Buildings listed in green are under construction and have topped out.
Buildings listed in red are under construction, but have not yet topped out.

Rank	Building Name	City	Year	Stories	Height m	Height ft	Material	Use
1	Burj Khalifa	Dubai (AE)	2010	163	828	2,717	steel / concrete	office / residential / hotel
	Ping An Finance Center	Shenzhen (CN)	2016	115	660	2,165	composite	office
	Wuhan Greenland Center	Wuhan (CN)	2017	125	636	2,087	composite	hotel / residential / office
	Shanghai Tower	Shanghai (CN)	2014	121	632	2,073	composite	hotel / office
2	Makkah Royal Clock Tower Hotel	Mecca (SA)	2012	120	601	1,972	steel / concrete	other / hotel / multiple
	Goldin Finance 117	Tianjin (CN)	2015	117	597	1,957	composite	hotel / office
	Lotte World Tower	Seoul (KR)	2015	123	555	1,819	composite	hotel / office
	One World Trade Center	New York City (US)	2014	104	541	1,776	composite	office
	The CTF Guangzhou	Guangzhou (CN)	2017	111	530	1,739	composite	hotel / residential / office
	Tianjin Chow Tai Fook Binhai Center	Tianjin (CN)	2016	97	530	1,739	composite	residential / hotel / office
	Zhongguo Zun	Beijing (CN)	2016	108	528	1,732	–	office
	Busan Lotte Town Tower	Busan (KR)	2016	107	510	1,674	composite	residential / hotel / office
3	Taipei 101	Taipei (TW)	2004	101	508	1,667	composite	office
4	Shanghai World Financial Center	Shanghai (CN)	2008	101	492	1,614	composite	hotel / office
5	International Commerce Centre	Hong Kong (CN)	2010	108	484	1,588	composite	hotel / office
	International Commerce Center 1	Chongqing (CN)	2016	99	468	1,535	composite	hotel / office
	Tianjin R&F Guangdong Tower	Tianjin (CN)	2016	91	468	1,535	composite	residential / hotel / office
	Lakhta Center	St. Petersburg (RU)	2018	86	463	1,517	composite	office
	Riverview Plaza A1	Wuhan (CN)	2016	82	460	1,509	–	hotel / office
6	Petronas Tower 1	Kuala Lumpur (MY)	1998	88	452	1,483	composite	office
6	Petronas Tower 2	Kuala Lumpur (MY)	1998	88	452	1,483	composite	office
	Suzhou Supertower	Suzhou (CN)	2016	92	452	1,483	–	residential / hotel / office
8	Zifeng Tower	Nanjing (CN)	2010	66	450	1,476	composite	hotel / office
9	Willis Tower	Chicago (US)	1974	108	442	1,451	steel	office
	World One	Mumbai (IN)	2015	117	442	1,450	composite	residential
10	KK100	Shenzhen (CN)	2011	100	442	1,449	composite	hotel / office
11	Guangzhou International Finance Center	Guangzhou (CN)	2010	103	439	1,439	composite	hotel / office
	Wuhan Center	Wuhan (CN)	2015	88	438	1,437	composite	hotel / residential / office
	Marina 101	Dubai (AE)	2014	101	432	1,417	concrete	residential / hotel
	Diamond Tower	Jeddah (SA)	–	93	432	1,417	–	residential
	432 Park Avenue	New York City (US)	2015	89	426	1,398	concrete	residential
12	Trump International Hotel & Tower	Chicago (US)	2009	98	423	1,389	concrete	residential / hotel
13	Jin Mao Building	Shanghai (CN)	1999	88	421	1,380	composite	hotel / office
14	Princess Tower	Dubai (AE)	2012	101	413	1,356	steel / concrete	residential
15	Al Hamra Tower	Kuwait City (KW)	2011	80	413	1,354	concrete	office
16	Two International Finance Centre	Hong Kong (CN)	2003	88	412	1,352	composite	office
	Huaguoyuan Tower 1	Guiyang (CN)	–	64	406	1,332	–	–
	Huaguoyuan Tower 2	Guiyang (CN)	–	64	406	1,332	–	–
	Nanjing Olympic Suning Tower	Nanjing (CN)	2016	89	400	1,312	–	residential / hotel / office
	Ningbo Center	Ningbo (CN)	2017	–	398*	1,306	–	hotel / residential / office
17	23 Marina	Dubai (AE)	2012	90	393	1,289	concrete	residential
18	CITIC Plaza	Guangzhou (CN)	1996	80	390	1,280	concrete	office
	Capital Market Authority Headquarters	Riyadh (SA)	2014	77	385	1,263	composite	office
19	Shun Hing Square	Shenzhen (CN)	1996	69	384	1,260	composite	office
	Eton Place Dalian Tower 1	Dalian (CN)	2014	80	383	1,257	composite	hotel / office
	Abu Dhabi Plaza	Astana (KZ)	2017	88	382	1,253	–	residential
	The Domain	Abu Dhabi (AE)	2013	88	381	1,251	concrete	residential
20	Empire State Building	New York City (US)	1931	102	381	1,250	steel	office
21	Elite Residence	Dubai (AE)	2012	87	380	1,248	concrete	residential
	Three World Trade Center	New York City (US)	–	71	378	1,240	composite	office

* estimated height

Rank	Building Name	City	Year	Stories	Height m	ft	Material	Use
	Gemdale Gangxia Tower 1	Shenzhen (CN)	2016	–	375	1,230	–	residential / office
22	Central Plaza	Hong Kong (CN)	1992	78	374	1,227	concrete	office
	Lamar Tower 1	Jeddah (SA)	2014	87	372	1,220	concrete	residential / office
	Oberoi Oasis Residential Tower	Mumbai (IN)	2015	82	372	1,220	concrete	residential
	The Address The BLVD	Dubai (AE)	2015	72	370	1,214	–	residential / hotel
23	Bank of China Tower	Hong Kong (CN)	1990	72	367	1,205	composite	office
24	Bank of America Tower	New York City (US)	2009	55	366	1,200	composite	office
	Dalian International Trade Center	Dalian (CN)	2015	86	365	1,199	composite	residential / office
	VietinBank Business Center Office Tower	Hanoi (VN)	2016	68	363	1,191	composite	office
25	Almas Tower	Dubai (AE)	2008	68	360	1,181	concrete	office
25	The Pinnacle	Guangzhou (CN)	2012	60	360	1,181	concrete	office
	Federation Towers – Vostok Tower	Moscow (RU)	2014	93	360	1,181	concrete	residential / hotel / office
27	JW Marriott Marquis Hotel Dubai Tower 1	Dubai (AE)	2012	82	355	1,166	concrete	hotel
27	JW Marriott Marquis Hotel Dubai Tower 2	Dubai (AE)	2013	82	355	1,166	concrete	hotel
29	Emirates Tower One	Dubai (AE)	2000	54	355	1,163	composite	office
	Forum 66 Tower 2	Shenyang (CN)	2015	68	351	1,150	composite	office
30	Tuntex Sky Tower	Kaohsiung (TW)	1997	85	348	1,140	composite	hotel / office
31	Aon Center	Chicago (US)	1973	83	346	1,136	steel	office
32	The Center	Hong Kong (CN)	1998	73	346	1,135	steel	office
33	John Hancock Center	Chicago (US)	1969	100	344	1,128	steel	residential / office
	ADNOC Headquarters	Abu Dhabi (AE)	2014	76	342	1,122	concrete	office
	Ahmed Abdul Rahim Al Attar Tower	Dubai (AE)	2014	76	342	1,122	concrete	residential
	Xiamen International Centre	Xiamen (CN)	2016	61	340	1,115	composite	office
	The Wharf Times Square 1	Wuxi (CN)	2015	68	339	1,112	composite	hotel / office
	Chongqing World Financial Center	Chongqing (CN)	2014	73	339	1,112	composite	office
	Mercury City Tower	Moscow (RU)	2013	75	339	1,112	concrete	residential / office
	Four Seasons Tower	Tianjin (CN)	2015	65	338	1,109	composite	residential / hotel
	Orchid Crown Tower A	Mumbai (IN)	2015	75	337	1,106	concrete	residential
	Orchid Crown Tower B	Mumbai (IN)	2015	75	337	1,106	concrete	residential
	Orchid Crown Tower C	Mumbai (IN)	2015	75	337	1,106	concrete	residential
34	Tianjin Global Financial Center	Tianjin (CN)	2011	75	337	1,105	composite	office
34	The Torch	Dubai (AE)	2011	79	337	1,105	concrete	residential
36	Keangnam Hanoi Landmark Tower	Hanoi (VN)	2012	70	336	1,102	concrete	hotel / residential / office
	Oko Tower 1	Moscow (RU)	2015	91	336	1,101	concrete	residential / hotel
	DAMAC Residenze	Dubai (AE)	2016	86	335	1,099	steel / concrete	residential
37	Shimao International Plaza	Shanghai (CN)	2006	60	333	1,094	concrete	hotel / office
38	Rose Rayhaan by Rotana	Dubai (AE)	2007	71	333	1,093	composite	hotel
	Tianjin Kerry Center	Tianjin (CN)	2015	72	333	1,093	steel	office
	China Chuneng Tower	Shenzhen (CN)	2016	–	333	1,093	–	–
	Modern Media Center	Changzhou (CN)	2013	57	332	1,089	composite	office
39	Minsheng Bank Building	Wuhan (CN)	2008	68	331	1,086	steel	office
40	China World Tower	Beijing (CN)	2010	74	330	1,083	composite	hotel / office
	Gate of Kuwait Tower	Kuwait City (KW)	2015	80	330	1,083	concrete	hotel / office
	The Skyscraper	Dubai (AE)	–	66	330	1,083	–	office
	Ryugyong Hotel	Pyongyang (KP)	–	105	330	1,083	concrete	hotel / office
	Suning Plaza Tower 1	Zhenjiang (CN)	2016	77	330	1,082	–	–
	Hon Kwok City Center	Shenzhen (CN)	2015	80	329	1,081	composite	residential / office
41	Longxi International Hotel	Jiangyin (CN)	2011	74	328	1,076	composite	residential / hotel
	Nanjing World Trade Center Tower 1	Nanjing (CN)	2015	69	328	1,076	–	hotel / office
	Wuxi Suning Plaza 1	Wuxi (CN)	2014	68	328	1,076	composite	hotel / office
	Concord International Centre	Chongqing (CN)	2016	62	328	1,076	composite	hotel / office
	Al Yaqoub Tower	Dubai (AE)	2013	69	328	1,076	concrete	residential / hotel
42	The Index	Dubai (AE)	2010	80	326	1,070	concrete	residential / office
	The Landmark	Abu Dhabi (AE)	2013	72	324	1,063	concrete	residential / office
	Deji Plaza Phase 2	Nanjing (CN)	2013	62	324	1,063	composite	office
	Yantai Shimao No. 1 The Harbour	Yantai (CN)	2014	59	323	1,060	composite	residential / hotel / office
43	Q1 Tower	Gold Coast (AU)	2005	78	323	1,058	concrete	residential
44	Wenzhou Trade Center	Wenzhou (CN)	2011	68	322	1,056	concrete	hotel / office
45	Burj Al Arab Hotel	Dubai (AE)	1999	60	321	1,053	composite	hotel
46	Nina Tower	Hong Kong (CN)	2006	80	320	1,051	concrete	hotel / office

Rank	Building Name	City	Year	Stories	Height m	ft	Material	Use
	White Magnolia Plaza 1	Shanghai (CN)	2015	66	320	1,048	composite	office
47	Chrysler Building	New York City (US)	1930	77	319	1,046	steel	office
47	New York Times Tower	New York City (US)	2007	52	319	1,046	steel	office
	Zhujiang New City Tower	Guangzhou (CN)	2015	67	319	1,046	composite	office
	Runhua Global Center 1	Changzhou (CN)	2015	72	318	1,043	composite	office
	Jiuzhou International Tower	Nanning (CN)	2015	71	318	1,043	–	–
	Riverside Century Plaza Main Tower	Wuhu (CN)	2015	66	318	1,043	composite	hotel / office
	United International Mansion	Chongqing (CN)	2013	67	318	1,043	concrete	office
49	HHHR Tower	Dubai (AE)	2010	72	318	1,042	concrete	residential
50	Bank of America Plaza	Atlanta (US)	1993	55	317	1,040	composite	office
	Bashang Jie North Tower	Hefei (CN)	2015	–	317	1,040	–	–
	Wuhan Qiaokou Project 1	Wuhan (CN)	2016	63	315	1,033	–	–
	Youth Olympics Center Tower 1	Nanjing (CN)	2015	68	315	1,032	composite	–
	Maha Nakhon	Bangkok (TH)	2015	77	313	1,028	concrete	residential / hotel
	Yunrun International Tower	Huaiyin (CN)	2015	75	312	1,024	–	office
	The Stratford Residences	Makati (PH)	2015	74	312	1,024	concrete	residential
	Moi Center Tower A	Shenyang (CN)	2013	75	311	1,020	composite	hotel / office
51	U.S. Bank Tower	Los Angeles (US)	1990	73	310	1,018	steel	office
52	Ocean Heights	Dubai (AE)	2010	83	310	1,017	concrete	residential
52	Menara Telekom	Kuala Lumpur (MY)	2001	55	310	1,017	concrete	office
	Palais Royale	Mumbai (IN)	2014	88	310	1,017	concrete	residential
	The Wharf IFC	Chongqing (CN)	2016	–	310	1,017	–	–
	Bashang Jie South Tower	Hefei (CN)	2015	–	310	1,016	–	–
54	Pearl River Tower	Guangzhou (CN)	2012	71	309	1,015	composite	office
	Guangzhou Fortune Center	Guangzhou (CN)	2014	68	309	1,015	composite	office
55	Emirates Tower Two	Dubai (AE)	2000	56	309	1,014	concrete	hotel
	Guangfa Securities Headquarters	Guangzhou (CN)	2016	62	308	1,010	–	office
	Burj Rafal	Riyadh (SA)	2014	68	308	1,010	concrete	residential / hotel
56	Franklin Center – North Tower	Chicago (US)	1989	60	307	1,007	composite	office
	Lokhandwala Minerva	Mumbai (IN)	2015	83	307	1,007	concrete	residential
	Infinity Tower	Dubai (AE)	2013	76	306	1,005	concrete	residential
57	The Shard	London (GB)	2013	73	306	1,004	composite	residential / hotel / office
	One57	New York City (US)	2014	79	306	1,004	steel / concrete	residential / hotel
	East Pacific Center Tower A	Shenzhen (CN)	2013	85	306	1,004	concrete	residential
58	JPMorgan Chase Tower	Houston (US)	1982	75	305	1,002	composite	office
58	Etihad Towers T2	Abu Dhabi (AE)	2011	80	305	1,002	concrete	residential
60	Northeast Asia Trade Tower	Incheon (KR)	2011	68	305	1,001	composite	residential / hotel / office
	Eurasia	Moscow (RU)	2014	72	305	1,000	composite	residential / hotel / office
61	Baiyoke Tower II	Bangkok (TH)	1997	85	304	997	concrete	hotel
	Shenzhen World Finance Center	Shenzhen (CN)	2016	68	304	997	composite	office
	Wuxi Maoye City – Marriott Hotel	Wuxi (CN)	2013	68	304	997	composite	hotel
62	Two Prudential Plaza	Chicago (US)	1990	64	303	995	concrete	office
	Diwang International Fortune Center	Liuzhou (CN)	2014	75	303	994	composite	residential / hotel / office
	KAFD World Trade Center	Riyadh (SA)	2014	67	303	994	concrete	office
63	Leatop Plaza	Guangzhou (CN)	2012	64	303	993	composite	office
64	Wells Fargo Plaza	Houston (US)	1983	71	302	992	steel	office
64	Kingdom Centre	Riyadh (SA)	2002	41	302	992	steel / concrete	residential / hotel / office
66	The Address	Dubai (AE)	2008	63	302	991	concrete	residential / hotel
67	Capital City Moscow Tower	Moscow (RU)	2010	76	302	990	concrete	residential
	Gate of the Orient	Suzhou (CN)	2014	68	302	990	composite	residential / hotel / office
	Heung Kong Tower	Shenzhen (CN)	2014	70	301	987	composite	hotel / office
68	Doosan Haeundae We've the Zenith Tower A	Busan (KR)	2011	80	300	984	concrete	residential
68	Arraya Tower	Kuwait City (KW)	2009	60	300	984	concrete	office
68	Aspire Tower	Doha (QA)	2007	36	300	984	composite	hotel / office
	Supernova	Noida (IN)	2015	80	300	984	–	residential
	Greenland Center Tower 1	Zhengzhou (CN)	2016	78	300	984	composite	office
	Greenland Center Tower 2	Zhengzhou (CN)	2016	78	300	984	composite	office
	Dubai Pearl Tower	Dubai (AE)	2016	73	300	984	concrete	residential
	NBK Tower	Kuwait City (KW)	2014	70	300	984	concrete	office
	Namaste Tower	Mumbai (IN)	2015	62	300*	984	concrete	hotel / office

Rank	Building Name	City	Year	Stories	Height m	Height ft	Material	Use
	Shenglong Global Center	Fuzhou (CN)	2016	57	300	984	–	office
	Kingkey Xiasha Project	Shenzhen (CN)	–	–	300	984	–	–
	Jin Wan Plaza 1	Tianjin (CN)	2015	66	300	984	–	hotel / office
	Abeno Harukas	Osaka (JP)	2014	62	300	984	steel	hotel / office / retail
	Gran Torre Costanera	Santiago (CL)	2013	60	300	984	concrete	office
	Langham Hotel Tower	Dalian (CN)	2015	74	300	983	–	residential / hotel
71	First Bank Tower	Toronto (CA)	1975	72	298	978	steel	office
71	One Island East	Hong Kong (CN)	2008	68	298	978	concrete	office
	Yujiabao Administrative Services Center	Tianjin (CN)	2015	60	298	978	–	office
	Four World Trade Center	New York City (US)	2013	64	298	977	composite	office
73	Eureka Tower	Melbourne (AU)	2006	91	297	975	concrete	residential
74	Comcast Center	Philadelphia (US)	2008	57	297	974	composite	office
	Dacheng Financial Business Center Tower A	Kunming (CN)	2015	–	297	974	steel	hotel / office
75	Landmark Tower	Yokohama (JP)	1993	73	296	972	steel	hotel / office
	Park Hyatt Guangzhou	Guangzhou (CN)	2013	66	296	972	composite	residential / hotel / office
76	Emirates Crown	Dubai (AE)	2008	63	296	971	concrete	residential
	Xiamen Shimao Cross–Strait Plaza Tower B	Xiamen (CN)	2015	67	295	969	–	office
77	Khalid Al Attar Tower 2	Dubai (AE)	2011	66	294	965	concrete	hotel
78	311 South Wacker Drive	Chicago (US)	1990	65	293	961	concrete	office
	Lamar Tower 2	Jeddah (SA)	2014	84	293	961	concrete	residential / office
	Greenland Puli Center	Jinan (CN)	2015	61	293	960	composite	residential / office
79	Sky Tower	Abu Dhabi (AE)	2010	74	292	959	concrete	residential / office
80	Haeundae I Park Marina Tower 2	Busan (KR)	2011	72	292	958	composite	residential
81	SEG Plaza	Shenzhen (CN)	2000	71	292	957	concrete	hotel / office
	Indiabulls Sky Suites	Mumbai (IN)	2015	75	291	955	concrete	residential
82	70 Pine Street	New York City (US)	1932	67	290	952	steel	office
	Hunter Douglas International Plaza	Guiyang (CN)	2014	69	290	951	composite	office
	Powerlong Center Tower 1	Tianjin (CN)	2014	59	290	951	composite	office
	Busan IFC Landmark Tower	Busan (KR)	2014	63	289	948	–	office
	Jiangxi Nanchang Greenland Central Plaza 1	Nanchang (CN)	2014	59	289	948	composite	office
	Jiangxi Nanchang Greenland Central Plaza 2	Nanchang (CN)	2014	59	289	948	composite	office
	Dongguan TBA Tower	Dongguan (CN)	2013	68	289	948	composite	hotel / office
83	Key Tower	Cleveland (US)	1991	57	289	947	composite	office
84	Shaoxing Shimao Crown Plaza	Shaoxing (CN)	2012	60	288	946	composite	hotel / office
85	Plaza 66	Shanghai (CN)	2001	66	288	945	concrete	office
85	One Liberty Place	Philadelphia (US)	1987	61	288	945	steel	office
85	Yingli International Finance Centre	Chongqing (CN)	2012	58	288	945	concrete	office
	Kaisa Center	Huizhou (CN)	2014	66	288	945	composite	hotel / office
	Soochow International Plaza East Tower	Huzhou (CN)	2014	–	288	945	composite	hotel / office
	Soochow International Plaza West Tower	Huzhou (CN)	2014	–	288	945	composite	residential
	Chongqing Poly Tower	Chongqing (CN)	2013	58	287	941	concrete	office / hotel
88	Millennium Tower	Dubai (AE)	2006	59	285	935	concrete	residential
88	Sulafa Tower	Dubai (AE)	2010	75	285	935	concrete	residential
90	Tomorrow Square	Shanghai (CN)	2003	60	285	934	concrete	residential / hotel / office
91	Columbia Center	Seattle (US)	1984	76	284	933	composite	office
92	Trump Ocean Club International Hotel	Panama City (PA)	2011	70	284	932	concrete	residential / hotel
92	Three International Finance Center	Seoul (KR)	2012	55	284	932	composite	office
	D1 Tower	Dubai (AE)	2013	80	284	932	concrete	residential
94	Chongqing World Trade Center	Chongqing (CN)	2005	60	283	929	concrete	office
95	Cheung Kong Centre	Hong Kong (CN)	1999	63	283	928	steel	office
96	The Trump Building	New York City (US)	1930	71	283	927	steel	office
97	Suzhou RunHua Global Building A	Suzhou (CN)	2010	49	282	925	composite	office
	Al Hekma Tower	Dubai (AE)	2014	64	282	925	steel / concrete	office
98	Doosan Haeundae We've the Zenith Tower B	Busan (KR)	2011	75	282	924	concrete	residential
	Indiabulls Sky Forest	Mumbai (IN)	2015	80	281	922	concrete	residential
99	Bank of America Plaza	Dallas (US)	1985	72	281	921	composite	office
99	Torre Vitri	Panama City (PA)	2012	75	281	921	concrete	residential

CTBUH Organizational Members
(as of April, 2013)

(List continued on next page)

T.Y. Lin International Pte. Ltd.
Tongji Architectural Design (Group) Co., Ltd.
Walter P. Moore and Associates, Inc.
Werner Voss + Partner
Yolles

Contributors:

Aedas
Allford Hall Monaghan Morris Ltd.
Alvine Engineering
Barker Mohandas, LLC
Bates Smart
Benoy Limited
Bonacci Group
Boundary Layer Wind Tunnel Laboratory
Bouygues Construction
The British Land Company PLC
C S Structural Engineering, Inc.
Canary Wharf Group, PLC
Canderel Management, Inc.
CBRE Group, Inc.
CCL Qatar w.l.l.
Continental Automated Buildings Association (CABA)
DBI Design Pty Ltd
DCA Architects Pte Ltd
Deerns Consulting Engineers
DHV Bouw en Industrie
DK Infrastructure Pvt. Ltd.
DongYang Structural Engineers Co., Ltd.
Ellumus LLC
Far East Aluminium Works Co., Ltd.
Goettsch Partners
Graziani + Corazza Architects Inc.
HAEAHN Architecture, Inc.
Hariri Pontarini Architects
Hiranandani Group
Hyder Consulting (Shanghai)
Israeli Association of Construction and Infrastructure Engineers (IACIE)
J. J. Pan and Partners, Architects and Planners
Jiang Architects & Engineers
KHP Konig und Heunisch Planungsgesellschaft
Langdon & Seah Singapore
Lend Lease
Liberty Group Properties
M Moser Associates Ltd.
Mori Building Co., Ltd.
Mutua Madrilena
Nabih Youssef & Associates
National Fire Protection Association
National Institute of Standards and Technology (NIST)
Norman Disney & Young
The Ornamental Metal Institute of New York
Otis Elevator Company
Parsons Brinckerhoff
Perkins + Will
Pomeroy Studio Pte Ltd
PositivEnergy Practice, LLC
RAW Design Inc.
Rosenwasser/Grossman Consulting Engineers, PC
Samcon Gestion Inc.
SAMOO Architects & Engineers
Sanni, Ojo & Partners
Silvercup Studios
SilverEdge Systems Software, Inc.
SIP Project Managers Pty Ltd
The Steel Institute of New York
Tekla Corporation
Terrell Group
ThyssenKrupp Elevator
TSNIIEP for Residential and Public Buildings
University of Illinois at Urbana-Champaign
Vetrocare SRL
Wilkinson Eyre Architects

Participants:

ACSI (Ayling Consulting Services Inc)
Aidea Philippines, Inc.
AKF Group, LLC
Al Jazera Consultants
ALT Cladding, Inc.
ARC Studio Architecture + Urbanism
ArcelorMittal
Architects 61 Pte., Ltd.
Architectural Design & Research Institute of Tsinghua University
Architectural Institute of Korea
Architectus

Arquitectonica International Corp.
ASL Sencorp
Atkins
Azrieli Group Ltd.
Bakkala Consulting Engineers Limited
BAUM Architects
BDSP Partnership
Beca Group
Benchmark
BG&E Pty., Ltd.
BIAD (Beijing Institute of Architectural Design)
Bigen Africa Services (Pty) Ltd.
Billings Design Associates, Ltd.
bKL Architecture LLC
BluEnt
Boston Properties, Inc.
Broadway Malyan
Callison, LLC
Camara Consultores Arquitectura e Ingeniería
Capital Group
Case Foundation Co.
CB Engineers
CCHRB (Chicago Committee on High-Rise Buildings)
CDC Curtain Wall Design & Consulting, Inc.
Central Scientific and Research Institute of Engineering Structures "SRC Construction"
China Academy of Building Research
China Institute of Building Standard Design & Research (CIBSDR)
Chinachem Group
City Developments Limited
Code Consultants, Inc.
COOKFOX Architects
Cosentini Associates
COWI A/S
Cox Architecture Pty. Ltd.
CPP, Inc.
CS Associates, Inc.
CTL Group
Cundall
Dam & Partners Architecten
Dar Al-Handasah (Shair & Partners)
Degraeuwe Consulting
Delft University of Technology
Dennis Lau & Ng Chun Man Architects & Engineers (HK), Ltd.
dhk Architects Pty., Ltd.
Diar Consult
DSP Design Associates Pvt., Ltd.
Dunbar & Boardman
Edgett Williams Consulting Group, Inc.
Edmonds International USA
Eight Partnership Ltd.
Electra Construction LTD
ENAR, Envolventes Arquitectonicas
Ennead Architects LLP
Environmental Systems Design, Inc.
Epstein
Façade India Testing Inc.
Fortune Shepler Consulting
FXFOWLE Architects, LLP
Gale International / New Songdo International City Development, LLC
GCAQ Ingenieros Civiles S.A.C.
GEO Global Engineering Consultants
M/s. Glass Wall Systems (India) Pvt. Ltd
Gold Coast City Council
Gorproject (Urban Planning Institute of Residential and Public Buildings)
Grace Construction Products
Greyling Insurance Brokerage
Grimshaw Architects
Guangzhou Scientific Computing Consultants Co., Ltd.
GVK Elevator Consulting Services, Inc.
Halvorson and Partners
Haynes-Whaley Associates, Inc.
Heller Manus Architects
Henning Larsen Architects
Hilson Moran Partnership, Ltd.
Hong Kong Housing Authority
BSE, The Hong Kong Polytechnic University
Housing and Development Board
IECA Internacional S.A.
ingenhoven architects
Institute BelNIIS, RUE
INTEMAC, SA
Irwinconsult Pty., Ltd.
Iv-Consult b.v.
Jahn, LLC
Jaros Baum & Bolles
Jaspers-Eyers Architects
JBA Consulting Engineers, Inc.
JCE Structural Engineering Group, Inc.
JMB Realty Corporation

John Portman & Associates, Inc.
Johnson Pilton Walker Pty. Ltd.
JV "Alexandrov-Passage" LLC
Kalpataru Limited
KEO International Consultants
Kinetica Dynamics Inc.
King Saud University College of Architecture & Planning
KPFF Consulting Engineers
KPMB Architects
LCL Builds Corporation
Leigh & Orange, Ltd.
Lerch Bates, Inc.
Lerch Bates, Ltd. Europe
Living Architecture Inc.
Lobby Agency
Louie International Structural Engineers
Mace Limited
Magellan Development Group, LLC
Manitoba Hydro
Margolin Bros. Engineering & Consulting, Ltd.
James McHugh Construction Co.
McNamara / Salvia, Inc.
Michael Blades & Associates
New World Development Company Limited
Nikken Sekkei, Ltd.
Novawest LLC
NPO SODIS
O'Connor Sutton Cronin
Option One International, WLL
P&T Group
Palafox Associates
Paragon International Insurance Brokers Ltd.
Pelli Clarke Pelli Architects
PLP Architecture
PPG Industries, Inc.
Profica Project Management
Project and Design Research Institute "Novosibirsky Promstroyproject"
Rafael Viñoly Architects, PC
Read Jones Christoffersen Ltd.
Rene Lagos Engineers
Riggio / Boron, Ltd.
Ronald Lu & Partners
Roosevelt University – Marshall Bennett Institute of Real Estate
Sauerbruch Hutton Verwaltungsges mbH
schlaich bergermann und partner
Schock USA Inc.
Sematic SPA
Shimizu Corporation
SKS Associates
SMDP, LLC
SmithGroup
Southern Land Development Co., Ltd.
St. Francis Square Development Corp.
Stanley D. Lindsey & Associates, Ltd.
Stauch Vorster Architects
Stephan Reinke Architects, Ltd.
Studio Altieri S.p.A.
Sufrin Group
Taisei Corporation
Takenaka Corporation
Tameer Holding Investment LLC
Tandem Architects (2001) Co., Ltd.
Taylor Thomson Whitting Pty., Ltd.
TFP Farrells, Ltd.
Thermafiber, Inc.
Transsolar
The Trump Organization
Tyréns
University of Maryland – Architecture Library
University of Nottingham
UralNIIProject RAACS
Vanguard Realty Pvt., Ltd.
VDA (Van Deusen & Associates)
Vipac Engineers & Scientists, Ltd.
VOA Associates, Inc.
Walsh Construction Company
Weiss Architects, LLC
Werner Sobek Stuttgart GmbH & Co., KG
wh-p GmbH Beratende Ingenieure
Windtech Consultants Pty., Ltd.
WOHA Architects Pte., Ltd.
Wong & Ouyang (HK), Ltd.
Wordsearch
World Academy of Science for Complex Safety
WTM Engineers International GmbH
WZMH Architects
Y. A. Yashar Architects
Ziegler Cooper Architects

Supporting Contributors are those who contribute $10,000; Patrons: $6,000; Donors: $3,000; Contributors: $1,500; Participants: $750.